£15 95

Environment, Society, and Rural Change in Latin America

The past, present, and future in the countryside

Edited by
David A. Preston

JOHN WILEY & SONS
Chichester · New York · Brisbane · Toronto

Copyright © 1980 by John Wiley & Sons Ltd.

All rights reserved.

No part of this book may be reproduced by any means, nor transmitted, nor translated into a machine language without the written permission of the publisher.

British Library Cataloguing in Publication Data:

Environment, society, and rural change in Latin
   America.
   1. Latin America—Rural conditions
   I. Preston, David Anthony
   309.2'63'098   HN110.5   79-41481

   ISBN 0 471 27713 4

Photosetting by Thomson Press (India) Limited, New Delhi and printed in the United States of America

Contributors

David Preston	School of Geography, University of Leeds, England
Norman Long	Department of Anthropology, University of Durham, England
Cynthia Hewitt de Alcántara	Mexico City
Claude Collin Delavaud	Institut des Hautes Etudes de l'Amérique Latine, University of Paris, France
Michael Redclift	Institute of Latin American studies and Wye College, University of London, England
Anne Collin Delavaud	Institut des Hautes Etudes de l'Amérique Latine, University of Paris, France
Jean Revel-Mouroz	Institut des Hautes Etudes de l'Amérique Latine, University of Paris, France
Raymond Pebayle	Department of Geography, Université de Bretagne Occidentale, Brest, France
Lourdes Arizpe	Centro de Estudios Sociológicos, El Colegio de Mexico, Mexico
Steven Webster	Department of Anthropology, University of Auckland, New Zealand
Richard Wilkie	Department of Geography, University of Massachusetts, Amherst, Mass., USA
Hélène Rivière d'Arc	Institut des Hautes Etudes de l'Amérique Latine, University of Paris, France
Ernest Feder	Berghof Stiftung Für konfliktforschung, Freie Universität, Berlin.

Contents

Preface ix

List of Illustrations xi

1. Introduction
   David A. Preston 1

LAND REFORM

2. Land reform, livelihood, and power in rural Mexico
   Cynthia Hewitt de Alcántara 21

3. Agrarian reform in Peru
   Claude Collin Delavaud 37

4. Agrarian reform and rural change in Ecuador
   Michael R. Redclift and David A. Preston 53

COLONIZATION

5. From colonization to agricultural development: the case of coastal Ecuador
   Anne Collin Delavaud 67

6. Mexican colonization experience in the humid tropics
   Jean Revel-Mouroz.. 83

7. Rural innovation and the organization of space in southern Brazil
   Raymond Pebayle 103

CULTURAL CHANGE

8. Cultural change and ethnicity in rural Mexico
   Lourdes Arizpe 123

9. **Ethnicity in the southern Peruvian highlands**
   Steven S. Webster 135

RURAL EMIGRATION

10. **Migration and population imbalance in the settlement hierarchy of Argentina**
    Richard W. Wilkie 157

11. **Change and rural emigration in central Mexico**
    Hélène Rivière d'Arc 185

12. **Rural emigration and the future of agriculture in Ecuador**
    David A. Preston 195

EXTERNAL PENETRATION

13. **The new agrarian and agricultural change trends in Latin America**
    Ernest Feder 211

CONCLUSION

14. **Some concluding comments: directive change and the question of participation**
    Norman Long 235

Author index 249

Subject index 000

Preface

The rural population of Latin America is occupying a gradually less important role as the proportion of the continental population that lives in urban areas increases. Decision makers and government in general is urban based and often urban biased. This collection of essays intends to bring to the reader's attention a series of issues that are important now and that will continue to be so in the future. They are issues that concern rural people themselves and those of us whose working life has been spent studying them and sharing their life. This book attempts to encourage interest in the analysis of the problems of rural areas in Latin America by examining theory concerning rural development, by reporting on situations in different parts of the continent and by polemical discussion. We feel that there is a need for both information and thoughtful analysis of these problems. The authors are united by their concern for the rural people with whom they have worked; they come from a variety of disciplines and it would not have been appropriate to attempt to force each author to conform to a single pattern in his or her treatment of a specific topic. The variety of approaches should enrich the reader's understanding of the issues that we confront and likewise the range of national intellectual traditions of the contributors is a valuable source of variety. The essays for this volume were all written especially for it and were completed during 1978.

Translation of the articles from French into English was undertaken by Sally Bramham, Rosemary Preston, Philippe Chesters, and myself. The maps were largely drawn by John Dixon of the School of Geography of Leeds University where Sue Hughes and Margaret Hodgson typed different parts of the manuscript, . I am greatly indebted to the contributors who supported the idea of such a collection of essays, who tried to meet deadlines but who put up with the inevitable delays. They provided myself and Norman Long with a stimulating series of essays to read, consider, and comment on in our respective introductory and concluding essays.

We mark our concern for colleagues and friends in Latin America by dedicating this book to those who have worked in universities in Latin America and who have suffered on account of their beliefs. The majority of the authors' royalties from this volume will be given to World University Service to assist its work on behalf of students and academics in Latin America.

WORLD UNIVERSITY SERVICE

World University Service is an international non-governmental university organization of students and staff on campuses in Africa, the Americas, Asia, and Europe. It has consultative status with several of the United Nations' specialized agencies. WUS resists all infringement of freedom of thought and expression in study, research, and teaching; opposes all forms of discrimination, exploitation, and injustice, especially in education; and promotes university involvement alongside the less privileged in the solution of basic problems. WUS (UK) is an educational charity, one of 33 national committees affiliated to International WUS. Since the 1920s, WUS (UK) has helped many of the casualties of repression who have arrived in Britain as refugees. In recent years, WUS (UK) has assisted refugees from Chile, Uganda, and Ethiopia, and Kurdish students from Iraq, in the United Kingdom, and Zimbabwean and Palestinian students studying in their own country.

In 1974 WUS (UK) started a scholarship programme for refugee students and academics from Chile, with funds from the Overseas Development Ministry. Over 900 awards have been offered, for courses in subjects which are relevant and appropriate to development in Latin America, at universities in the United Kingdom. Awardholders are given special orientation and English language courses, and legal and welfare advice. Since 1977, WUS (UK) has developed a reorientation programme to encourage the coordination of studies and research among Chilean students and academics in the United Kingdom, and to assist refugees who have finished their studies to return to the region (if not to their own country) to contribute their skills to development. WUS (UK) works with other refugee organizations to improve the treatment of refugees internationally and in the United Kingdom.

World University Service (UK)
20–21 Compton Terrace
London N1
United Kingdom

List of Illustrations

Figure 5.1 Agricultural development in coastal Ecuador.
Figure 6.1 Mexico: growth of the cultivated area, 1950–73.
Figure 6.2 Mexico: active pioneer zones.
Figure 6.3 Colonization and development in the Grijalva and Usumacinta basins.
Figure 7.1 Southern Brazil.
Figure 9.1 The Cuzco region of Peru.
Figure 10.1 Depopulation in Argentina.
Figure 10.2 Migration in Argentina. Ratio of out- to in-migration by province, 1947.
Figure 10.3 Migration in Argentina. Ratio of out- to in-migration by province, 1960.
Figure 10.4 Migration in Argentina. Ratio of out- to in-migration by province, 1970.
Figure 10.5 Ratio of out- to in-migration in Argentina, 1947–70.
Figure 10.6 Migration volume in to and out of Aldea San Francisco, 1900–77 (yearly mean for each five-year period).
Figure 11.1 The central highlands of Mexico.
Figure 12.1 Depopulation in Ecuador.

Environment, Society, and Rural Change in Latin America
Edited by D. A. Preston
© 1980 John Wiley & Sons Ltd.

CHAPTER 1

Introduction

DAVID A. PRESTON

The rural areas of Latin America in the 1980s are going to be very different from what they were ten or twenty years ago, and the changes are already very obvious to those who have worked in them for a number of years. The extent to which rural areas are conceived as having problems depends very much on where one lives and the interests that are served by such an analysis. Townsfolk look upon rural areas as declining regions with a limited future and low standards of living; country people may regard their problems not as being rural in origin but rather urban, arising from the unwillingness of city-based Governments to provide for the rural population the range of services available in the city. For economists and 'developers' imbued with a wish for and belief in modernization, the rural areas are like wilful children who will not grow up and do not contribute to the welfare of the family/nation as they should.

In this collection of essays, the reader is invited to share our analysis of a series of issues that we feel are important to rural people and which are likely to remain important for the remainder of the century. Rural areas are important in Latin America although they no longer contain a majority of the population. They provide food and industrial raw materials for Latin American and other nations and, above all, their people are among the most underprivileged in the continent. Even those who want nothing of these privileges are likely to be affected by demands on their land and by the conversion of their children to the ethics and beliefs of urban society through schooling. Rural areas are especially vulnerable to outside influences because they are geographically remote from centres of political power, are frequently poorly organized, especially when many people are absent part of the year, and are often divided with respect to their needs. Often the State has provided them with token benefits which favour some, but not all, sections of local society; benefits such as land reform, which leave the beneficiaries temporarily satisfied, the former estate owners embittered, and the remaining landless discouraged. One may even reach the stage where one asks the question why have not even more people left the land, as Roberts has recently suggested (Roberts, 1978).

In identifying the principal issues on which the essays were prepared, the main criteria were that issues should be likely to be of continuing or growing importance with relation to rural people, that they should identify issues which were frequently identified by rural families as concerning them deeply, and, finally, that they should be issues about which the social science community was chary of expressing views. Much of the interesting literature in some branches of social science—in particular, geography, rural sociology, cultural anthropology and, to a lesser extent, agricultural economics—is either strongly theoretical with little relation to reality or else so descriptive that relation to theory is either absent or made difficult by the inadequate definition of local terminology. The essays in this volume seek to provide both empirical observations and some related theoretical statements about a series of rural phenomena which occur in many parts of Latin America and which are part of the everyday experience of Latin American country people.

AGRARIAN REFORM

The widespread concern with agrarian reform of the 1950s and 1960s has not been continued with much enthusiasm into the 1970s. It was commonplace for social scientists and liberal commentators to identify the most glaring social injustice as the unequal division of land, especially good farming land, in such a way that the few rich and powerful families had access to most of the best land. This focus of attention culminated in the publication, in the late 1960s, of a series of detailed studies of land distribution and socio-economic development in a series of countries as yet unaffected by a profound agrarian reform.[1] These documented, on a scale never previously attempted, the agrarian situation, including major summaries of existing legislation, the examination of national-level rural development, and, through a series of case studies, the investigation of conditions in estates, plantations, and freeholding indian and mestizo communities. As a second stage, studies were carried out in three countries which had very different experience of agrarian reform: Mexico, Bolivia, and Venezuela. While the results of the first-stage work were rapidly and widely diffused, the second-stage studies were not equally successful, and only the Mexico study has been formally published, although some results of the Bolivian work have had limited circulation.

The experience of different Latin American countries with respect to the results of differently conceived agrarian reforms is extraordinarily rich, and yet land reform is increasingly seen as difficult to achieve by gradualistic legislation. In countries that have experienced reform, attention becomes increasingly focused on those who have not benefited from the reform procedure. The emergence of new and equally powerful groups which improve their position seemingly to the detriment of the poor highlights the need for a more complete restructuring of society. The one example of a socialist agrarian reform does not appear to have inspired emulation by the non-socialist Governments that now predominate in Latin America. This is due in part to the special set of agrarian conditions prevailing in Cuba which set that country apart from almost all other Latin American nations—the existence of a rural proletariat that annually suffered long periods of unemployment, the importance of the United States investment in commercial agricultural production, and the relatively high level of national socio-economic development—and also because the success of Cuban agrarian reforms were not particularly evident. Agricultural production—the crucial measure of success for many conservative critics—was not buoyant, and important tactical failures, such as the 'Ten Million Ton Harvest', obscured the real gains in rural welfare and dynamic rural and provincial development. That only Chile subsequently chose a socialist road to progress was of obvious importance, and the most strikingly innovative agrarian reform since the Cuban Revolution—that of Peru—owes relatively little in its inspiration to socialist examples, but is more individual in its conception.

Agrarian reform has become a less explosive issue in Latin America not because the need for land redistribution and expropriation of large landholdings is any less than before but because of the emergence of new defenders of the current situation—the beneficiaries of gradualistic reforms—and because of the recognition that agrarian reform is a great deal more than the reallocation of a single productive resource, but really means a redistribution of power, as Cynthia Hewitt de Alcántara indicates in her essay, since it is the social inequalities that urgently demand eradication. Furthermore, other rural changes have attracted attention away from agrarian reform—the increased use of high-yielding varieties of food grains, usually associated with a package of modern farming practices which have increased production but largely contributed to the enrichment of middle-sized or large farmers, the increasing importance of the intervention of multinational corporations in agriculture, and the whole range of anxieties aroused by the growth of agri-business. Indeed, each of the topics with which this book is concerned has taken away a little of the importance that agrarian reform once had in Latin American affairs. The ineffective nature of many so-called reforms, which did little more than toy with distribution of poverty, also encouraged disenchantment in a view of agrarian reform as the key element in any scenario of rural improvement.

Each of the three essays on agrarian reform examines a particular situation representing the experience of one nation and also the consequences of radically different agrarian reform measures. In this context, the geographical location of the countries is much less important than the experience they display of specific reform policies. Mexico has the longest and most varied experience of agrarian reform in Latin America, and it is worth noting that the Mexican Revolution predates the Russian Revolution. The indigenous organization of the re-occupation of alienated common lands by Zapata and his followers was succeeded by redistribution by the new post-Revolutionary State which disregarded the earlier, locally implemented reorganization. Until the Presidency of Cárdenas, reform measures, though enshrined in the Constitution, languished.

The new collective and non-collective *ejidos* created new organizations separate from traditional community organization which caused friction in areas where traditional authority predominated. The *ejidos* were effectively controlled by the Government credit organizations, and those collectives concentrating on commercial production were limited by technical advisers employed by Central and State Government in their freedom of action. Since Government priorities with relation to agricultural development have been for increases in commercial production by the private agricultural sector, the development of Mexican agriculture since 1940 has been marked by a rapid increase in production and economic power of large landholders and the inability of a growing portion of the peasantry to subsist on what their plots produce. A consequence that relates to a further group of essays is the migration from the countryside of a part of the population, both for short periods—thus making farming a part-time activity—and for long periods. The landless poor, who might once have been able to subsist on wages from occasional work on large properties or as sharecroppers for *ejido* members, were also forced to seek their fortunes elsewhere. The existence of post-Revolutionary large estates and of four million rural workers with either no land or not enough to support a family makes access to land a live issue, at least in the minds of the rural population.

In Peru, by contrast, the archaic land tenure system, characterized by large agro-industrial farms producing crops for export in the coastal valleys and extensive livestock-rearing ranches in the mountains, experienced little real change until the agrarian reform measures introduced by the Military Government in 1969. An important distinction in Peru must be drawn between the conditions in the rural areas of the coast and the sierra. While, on the coast, there was a readily identifiable proletariat who were wage earners on the large agro-industrial farms as well as a much poorer group of smallholders who depended on the commercial farms for occasional but remunerative work, in the mountains there was no such large, proletarianized group of estate workers. There were numerous peasant communities, often organized along traditional lines, but the benefits of a takeover of the estate lands were seen more as providing access to land than as a means of securing a cash income.

The results of the Peruvian reform are less easily traceable than in the case of Mexico, in part because the reform is barely a decade old and because a change in Government has resulted in a new Administration which is less identified with rural progress through land reform. The Peruvian reform was clearly directed towards an overhaul of Peruvian society which could result in the more effective integration of the rural population in the life of the nation. The special interest of the Peruvian agrarian reform experience lies in the emphasis given to collective farming, with a strong emphasis to the creation of large administrative units, particularly in the highlands. The reasons behind this policy are clear and logical: to provide for a more effective horizontal exchange of goods, and to allow the least-developed rural communities to experience contact with workers on smallholdings, on former estates, and from different ecological zones. The policy is fraught with many difficulties and does not so far seem to have been successful in making people from widely separated areas aware of common problems and appreciative of the value of cooperation to overcome them. As in Mexico, the power of the State to direct certain aspects of rural development is greater than before.

The case of Ecuador is quite different, for no effective agrarian reform programme has been initiated by recent Governments, although some forms of servitude have been abolished and a small amount of land redistribution has taken place. In this way, Ecuador may represent the majority of Latin American countries who have agrarian reform measures in the legal code, but no effective agrarian reform agency with a clear mandate, backed by adequate funds, to change the land tenure system. Because the changes that have taken place are neither obvious nor have attracted national publicity, and because an urgent need for further reforms seems to exist, it is tempting to overlook the changes that have taken place. Since Ecuador includes semi-arid and humid tropical lowlands and highlands, and an agricultural economy which produces bananas, cacao, and coffee for export, as well as a wide range of other crops for national consumption, there are many situations under which land redistribution might be necessary. The major analysis of Ecuadorian land tenure conditions in the 1960s (CIDA, 1964) concluded that unequal access to land was a major factor holding back production in parts of the agricultural sector of the economy, and that rural poverty could not be reduced without a major transfer of resources. Important changes have occurred in Ecuador in association with the revision of agrarian reform laws but even more as a result of the increasing penetration of market forces into agricultural areas in which production was only partly commercially oriented. The generally liberal nature of the Ecuadorian Military Government of the 1970s, plus the additional Government revenue from the new petroleum fields of the northern jungle, combined to cause large landowners to feel increasingly threatened by the possibility that the existing land reform might one day be implemented. The sale of land by landowners has thus contributed more to a reallocation of land than the actual agrarian reform, both in the Guayas basin and in areas of the highlands. The possibility of the implementation of reform did cause an unknown number of serfs and tenants to lose their livelihood, and only a small proportion of the rural poor derived any benefit whatsoever.

Agrarian reform has not been a wholly effective means of reducing income inequalities as a single development policy anywhere in Latin America. The greatest increase in rural well-being in comparison with urban standards has probably taken place in Cuba in the last twenty years, though much of this progress is not directly attributable to the creation of State farms but much more to an overall effort at rural planning and the massive provision of services in rural centres under State control. By contrast, State control in some other countries has been used to the detriment of the rural population in the implementation of policies which benefit urban areas and modern capitalist farming on larger-than-average-sized units, while the majority of the rural population and certainly the poorest rural people derive very little benefit. Indeed, it seems that the nature of agrarian reform is to benefit most those with direct access, however precarious, to some sort of estate land, and that landless casual labourers are effectively excluded by Government and the more secure rural workers from participation in agrarian reform programmes. The partial agrarian reforms that we have examined create a group of reform beneficiaries who may act as an interest group to block measures to benefit others less advantaged.

COLONIZATION

The settlement of new lands has had much more importance in the evolution of national identities and effective territories in Latin America than in many other developing areas because of the existence of relatively large underpopulated areas that are capable of supporting a larger population. Many Latin American countries have some rural areas with a high population density and a predominance of tiny farms that lie within one or two days walk from areas of adequate agricultural potential. Colonization is thus a recurrent theme in the history of settlement in much of Latin America, and yet it remains a topic shrouded in uncertainty and inadequate information. It is an important issue which relates to rural areas in the continent because colonization involves the creation of new rural societies and because the development of areas of new lands is thought to divert migrants from heading to major urban centres where inadequate housing and formal employment opportunities exist.

The development of colonization areas has been studied more by geographers, and yet there is little theory that has been developed in relation to it. Syntheses of existing studies are rare, and comparative studies are generally unsatisfactory (see Dozier, 1969), and yet the state of knowledge remains as inadequate as was suggested in a short review article written some seven years ago (Preston, 1974), although a series of conferences arranged by the Laboratoire Associé 111 in Paris is attempting to draw together current knowledge and develop some worthwhile generalizations. None of the essays on colonization in this volume pretends to make a significant contribution to theory related to land settlement, but each identifies characteristics of colonization in different parts of the continent and suggests a series of elements that are important in the pattern and process of land settlement.

As a summary judgment, one can suggest that the agricultural colonization of new land has not been remarkably successful in either generating a major surplus of goods for national or international trade or in providing a livelihood for a significant proportion of migrants. Spontaneous colonization does seem to be successful in affording farm families with a level of living and way of life that they find more congenial than that which they experienced in their previous homes. The major issues which concern the development of new settlements span the fields of social science disciplines and include important elements of man–environment relations encompassed by geographical interests. The nature of the new communities formed in areas of colonization, and their relationship to society as a whole, needs to be explored in order to discern whether such new communities are effectively a spatial extension of the communities from which migrants came, and also whether relationships between members of the locality group and between them and national society are structurally different from relationships in the urban and rural areas of origin. Roberts (1975) has recently commented that, in a case study in southern Colombia, the distribution of land and of power within new settlements mirrored closely that in highland Colombia, and Hélène Rivière d'Arc's essay in this volume and the work of Kleinpenning (1977) have shown that small farmers in Amazonian Brazil have little power in comparison with major landholders and financial interests from São Paulo and elsewhere. Yet colonization areas attract settlers, in part because of the fresh social opportunities they offer as well as because of the uncultivated land.

The variable nature of the environmental potential in areas of colonization has led to failure and impoverishment of many farmers and even of entire groups of colonists. Even given the nature of modern methods of land capability assessment, mistakes in the planning of colonization zones occur and, almost invariably, spontaneous colonization takes place with little benefit from land potential assessment and frequently in the less desirable areas, such as on hillsides above fertile land in river valleys which has already been settled. The issue related to this problem concerns the means of offering new settlers strategies whereby they can improve the quality of their life in the manner that they prefer and in harmony with the opportunities offered by the physical environment, so that the farming systems can offer a secure income and supply of food. This issue must necessarily embrace settlers of the various categories mentioned by Pebayle, namely those on worn-out land as well as those on steep slopes or with eroding soils who need both to restore the land and the vegetation cover and to provide themselves with food and clothing. These issues are clearly important in the cases examined by both Revel-Mouroz and Pebayle.

A third issue arises in relation to experience in both Mexico and coastal Ecuador as described by Revel-Mouroz and Anne Collin Delavaud. Colonization areas that are described in the literature and in these essays are very much part of national and international economic systems, and thus much development is associated with either the demand for a particular export crop or Government policies to promote a particular agricultural strategy. In her examination of the experience of colonization in coastal Ecuador, Anne Collin Delavaud emphasizes not only how farming developed in relation to the demand for a particular crop, as interpreted by some farmers and national Governments, but also how particular Government development policies favoured the development of some areas and some categories of farmers rather than others. The issue, therefore, is the extent to which the farming systems, including land tenure as well as the nature of the crops sown, really develop as a response to external pressures and not so much as a result of individual initiatives.

This refers only to the market-oriented colonization areas and in Ecuador, certainly, there are many areas of land settlement that are increasing in population but are very isolated from national markets, where Government influence is negligible and which, in highland Andean areas, may best be thought of as extensions of zones of influence of highland communities and a contemporary reassertion of influence over multiple ecological zones. Access to produce from a range of ecological zones—from high-altitude camelid pasture through potato and maize zones down to sugar-cane lands in the low-altitude humid tropical lowlands—seems to have been widespread in pre-Conquest Andean communities as John Murra and his students have described and as has been most elegantly demonstrated in Ecuador by Frank Salomon (1978). Many of the communities were gradually deprived of access to areas distant from the most permanent settlements—a process described by Favre with respect to communities near Huancavelica (Favre, 1977)—but increasing strains on the resource base has pushed some farmers to move regularly to lower altitudes, often several days walk away, where they clear land and sow crops which sustain these families during part of the year without any intention of providing a cash income. In some cases, lowland fields are rented but, in others, they are acquired and titles issued, and such lands become effectively part of the domestic production unit which comprises an archipelago of production zones.

This type of colonization, which is referred to by Revel-Mouroz in this volume and elsewhere, is less subject to Government plans to increase the production of a specific crop and is frequently described as an adjunct to a highland land use system as in the case of the Tzotzil (Collier, 1975) in Mexico and the Saraguros in Ecuador (Stewart et al., 1976).

State intervention in the form of planned settlement and the provision of powerful financial incentives, as is shown by Revel-Mouroz in Mexico, is not always successful, and errors in planning and execution of projects result in a continual failure to match planned targets, while the privileges acquired by some colonists may embitter others not so privileged and create considerable friction within newly evolving communities.

Finally, Pebayle's essay on recent experience in southern Brazil, in areas which have been farmed in one way or another for several generations, emphasizes that colonization is usefully visualized as a long process, that land experiences a succession of waves of colonization and use, and that it is a great deal more than the clearing and farming of new land, but includes the regeneration of or, at least, reappraisal of farmland and the development of new systems of land use, new patterns of urban growth, and a new relationship with national and regional economies. This series of essays relating to colonization thus demonstrates not only important issues concerning colonization but also that colonization as a phenomenon is subject to influences very similar to those that affect rural society in areas that have been longer settled.

SOCIAL CHANGE AND ETHNICITY

The original inhabitants of the Americas are now a tiny minority in the New World; they have disappeared almost completely from some countries and yet still may form a majority in others. The fate of lowland, forest indians has attracted wide interest and serious documentaries that portray the struggles of survivors to maintain their livelihood command large audiences, as in the 'Disappearing World' series of Granada Television—a commercial company—in the United Kingdom. Not all tribes have their documentary films and the indians of highland Bolivia, Peru, Ecuador, or Guatemala attract tourists, but there is little concern about their disappearance. Yet the cultural content of indian life is changing, and some young people from indian villages are leaving their community and living much as mestizos in colonization areas, city slums, or shanty towns.

The title of this section asks whether the gradual process of social change—modification in the patterns of behaviour and in relations between individuals, families, and communities—involves a loss of ethnicity, of belonging to a human group whose beliefs and values are different from other groups and which is expressed in specific clothes, behaviour, and frequently identified with a particular locality. In indian areas, the social distance—and often spatial distance too—between indian and non-indian is considerable, and the terms used to describe the two groups are specific and imply that nothing such as a mestizo-dressed indian or a downwardly mobile mestizo could exist. In the field, many areas exist where the rural population has a few items of indian culture and yet is not apparently indian and is unwilling to be so identified, and, of course, poverty is by no means confined to indigenous people.

What the essays in this section of the book are concerned with is the way in which ethnic affiliation may change and its relation to other social configurations and with what other factors these changes occur. A major debate among social scientists, although less lively now than previously, concerns the rôle of class in change, and one school of thought would suggest the pre-eminence of class and the relative unimportance of ethnic affiliation. A powerful argument in favour of such a view is the almost universal penetration of an urban-biased view of society which extols the virtues of the acquisition of manifold goods which are obtainable only by selling more than one did before of farm produce in order to live 'decently'. The influence of certain institutions or forms of activity on rural communities everywhere has been to introduce new, urban-biased ideas and values, although, as Webster emphasizes, some communities have diverse strategies of minimizing such influences. The introduction of schools, the gradually facilitated impact of radio providing screened information about other areas, and the increasing wave of rural emigration each contribute to introducing alien elements into local life, whether or not this affects community culture. What the essays in this volume are more concerned with are the ways in which local culture is modified and the extent to which this is also social, rather than cultural, change.

If the cultural distinctiveness of indians *is* gradually disappearing as blue jeans replace homespun trousers, factory beer replaces *chicha*, and brass bands oust wind ensembles, does it matter either to members of rural communities experiencing such change or to social scientists concerned with attempting to understand more about rural societies? If what is replacing former customs and values is quite alien and may lead to a decrease in the range of opportunities for adults and young to make a living, then surely such changes should be contested. If such change was clearly not in the interests of the participant, then it probably would not be adopted, and thus the question has little relevence. Many of the changes decrease the variety of goods, crops, or livestock available, but the efficacy of those that are available is increased and people are materially better off.

Where indian groups have disappeared within the past one or two generations (the Uru of Bolivia and Peru, and the Ona of Patagonia), should their disappearance be a cause for sorrow. Although, personally, I decry a decrease in the range of ethnic variety in a continental area, the cause for concern—in the case of the Ona, certainly—is far more often the means by which the tribe became extinct rather than the actual extinction of the tribe.

It is necessary to understand the strategy for survival of specific human groups faced with stress and to understand the process of change in order to know how to counsel those seeking help in both Government and communities, and to increase our knowledge of the means by which society adapts to new social and cultural situations. The two essays both use case-study material—Arizpe studies Mazahua villages and their urban migrant representatives, and Webster considers change and ethnicity in the context of an extremely isolated group of villages in southern Peru. There are, however, important differences in analysis. While Arizpe stresses the evolution of Mexican society and the emergence of various pressures on Mazahua indians, in particular economic forces, Webster is more concerned with the ways in which one particularly isolated community responds to change from outside, and he stresses strongly the heavy bias in evidence of social change in favour of less isolated places.

An important element of Arizpe's analysis focuses on the role of external forces in putting pressure on indians to identify to a greater extent with national society and become more Mexican than Mazahua, but she suggests that economic well-being both at a national and community level plays an important part in accelerating or holding back tendencies to 'modernize'. Thus, at times of national economic crisis, in indian communities with a relatively stable economic base, the tendency may be to have little contact with the national society and economy, and for greater economic and cultural self-sufficiency. Crisis within the community, however—repeated crop failure or shortage of land—may stimulate migration and increasing involvement with alien communities. Such a summary inevitably distorts a sensitive analysis, for the essence of the essay is to stress the range of reactions to particular situations and the many ways in which indians react to the need to earn a living outside their village and seek to retain their separate ethnic identity. The essay also identifies important changes in Mexican popular culture, now strongly influenced by the United States, which create an even greater cultural gap between the behaviour and values of rural indian and urban Mexican environments. The Mazahua communities have had contact with Mexico City and other migration destinations for several generations and are actively interacting with Mexican society. They are not necessarily representative of all indian communities, many of which are further from major cities and interacting much less intensely with non-indian society. Webster suggests that analysis of indian acculturation may easily be distorted through a tendency of social scientists to work in communities close to highways which are not very culturally isolated. Even so, some Mazahua villages were more closed than others and differential mestizo and indian response to change, in the same geographical area, is illuminating.

A series of villages on the eastern side of the Andes situated far from any highway and with only limited and infrequent contact with mestizo Peruvian society provides case-study material for Webster's essay. He stresses the power of communities in such situations to develop adaptations within the village without a need to have recourse to further contact with mestizos. In the indian community, sufficient stratification occurs—although unrecognized by non-indian outsiders—to provide stimulus for those wishing social mobility, and the institutions that bridge the ethnic gap do so precariously without necessarily forcing indian submission or establishing non-indian dominance, except in specific spheres. Thus, while mestizos may boast of indian exploitation, in some cases this is based upon premises that are not recognized by indians. Associated with this is the unconscious bias of an alien non-indian anthropologist whose views are confused and overlapping according to the social and ethnic identity of the informant. Webster argues that inequalities, which clearly exist in southern highland Peruvian society, are most satisfactorily identified along ethnic lines, although class differences can also be distinguished. Although it is possible to argue that Webster's highland community is a special case by reason of its isolation, many other such communities exist, and an understanding of the basis of social and ethnic relations in such a case adds an important perspective to any consideration of ethnicity in Latin America.

RURAL EMIGRATION

The migration of rural people to urban places and to other rural areas is commonplace in much of the contemporary world, but never ceases to attract comment. Action that controls this migration is unusual and rarely effective. It manifests itself throughout Latin America and occurs at the same time as the proportion of the population living in urban areas increases. The general awareness of the existence of widespread rural emigration and a feeling among Governments that it is undesirable—because many rural migrants head for metropolitan centres where severe overcrowding and widespread underemployment exists—tends to cloud appreciation of the nature of this population movement.

The movement of the rural population is one of long standing. It involves a large amount of return-migration, and the life experience of individuals usually comprises movement away from home to more than one destination and includes stays away of different lengths of time. Thus, comments on migration involve generalizations about complex experiences, and the interpretation of migration involves recording views of individuals looking back across both time and space and are thus not necessarily completely objective. Rivière d'Arc indicates in this volume, in line with many other analysts, that migration is associated with the economic situation in both origin and destination zone. Yet this situation stimulates only some people to move, and a more profound individualistic interpretation of migration might lay emphasis on the psycho-social situation of migrants where they feel aware of their inability to direct their life in the way they wish to go, while living in the country. They know that others like them have moved and achieved a more satisfactory life away, and they also feel that there are friends and kinsfolk where they are going who can offer care, counselling, and economic help to enable them to become established.

Two further characteristics of rural emigration which were identified as being frequently overlooked were its relative antiquity and the frequency of migrants returning. Although, in the past, there have been movements from urban centres to the rural areas when city life became unbearable as a result of industrial or commercial decline or civil war, some of the rural population have long since gone in search of a new life elsewhere, and it was such movements that in part provided settlers for the New World. At various times during the colonial period, Governments sought to inhibit rural indians from moving to the towns and, in other places, indians sought respite from extortion in migration far beyond Spanish control in isolated rural areas, as Browning describes for El Salvador (Browning, 1971, p. 117).

Return-migration has always been an integral part of the migration process and it should not therefore be surprising that people with previous migration experience form an important part of many rural areas, as the studies here of villages in Argentina and Ecuador reveal. This has important connotations with regard to the effect of migration on sending communities, since they are affected both by migrants leaving and by former migrants returning, and the interrelations of this ebb and flow of people are self-evident.

The three situations examined by Rivière d'Arc, Wilkie, and myself are analysed at different levels, but striking similarities emerge despite the difference in the cultural environment and the settlement history of each area. This, at once, suggests the importance of international trends which influence the modernization processes of many countries. As a consequence, farmers and rural people throughout the world feel deprived if they do not have access to the goods which others find indispensable.

The situation in Central Mexico is special because of the proximity of a major conurbation, large even on a world scale, and the importance in many parts of northern and central Mexico of migration to the United States. Even so, Rivière d'Arc's brief survey suggests the importance of factors that are not confined to Mexico. Although she stresses economic motives as a powerful impulse to migration, it is clear that all sorts of people migrate, and that the perception of economic weakness is by no means confined to the poorest people or even those in the poorest communities. The pace of culture change has also accelerated and has been associated with migration, as Arizpe indicates elsewhere in this volume. Major variations do occur both in the impact of emigration on sending communities and in levels of cultural change, and Arizpe analyses reasons underlying these differences which thereby illuminate both the processes that are transforming rural society.

Wilkie develops his analysis of migration to demonstrate perceptively not only the development of migration in the particular part of Entre Ríos, to which his field experience refers, but in relation to Argentinian society as a whole. He shows that even in a country whose population has grown within living memory as a result of international migration, the tide of internal migration towards Buenos Aires has developed, just as Mexico City has become progressively more important within Mexico as a migration destination. A possible cause of this change may well be the lack of dynamism of lower-order urban centres, but Wilkie also shows that the increased mobility of the rural population has permitted the use of services in larger, more-distant urban centres than formerly. The strong preference shown by informants for larger centres suggests that a strategy for urban development promotion has to take account of the viability of urban centres of a certain size.

At a different level of analysis, Wilkie is optimistic concerning the changes that have affected the Argentinian village as a result of migration, while my own experience in villages in highland Ecuador is more pessimistic regarding the nature of change. The emphasis given by rural families to the improvement in the quality of their life and the use of migrant savings to this end should not be surprising, but the slight emphasis given in Ecuadorian case-study areas to investment in agriculture to increase income is seen as a characteristic of these sorts of communities and not necessarily of all rural communities. Here, agriculture has long been regarded as being directed towards providing food rather than producing a surplus for sale. Cottage industries provided cash from the sale of manufactured goods—straw hats, ropes, blankets, etc.—until factory-made produce undermined their profitability. The intensification of agriculture is seen, in highland Ecuador, as difficult to achieve, and it is frequently middle-scale farmers who have adopted new crops and become commercial producers. The Ecuadorian historical experience of migration mirrors closely that in Argentina and that referred to by Rivière d'Arc in Mexico, with a gradual swing of migration from rural areas to the cities, rather than to other rural areas, in the past fifteen years.

Several issues arise from these three papers on rural emigration which relate to this book's central concern with the rural environment. In the first place, rural emigration is resulting at a national level in net transfers of population from rural areas to urban centres, although this is frequently partly obscured by the natural increase in population. Thus, many rural areas are actually losing population, houses are standing empty, and some land is being left uncultivated. Data from two such dissimilar countries as Argentina and Ecuador show the areas of each nation in which population is declining. Other evidence suggests that the rate of out-migration does more than just release what some have called population pressure. The issue associated with this empirical comment is the nature of the causes of rural emigration if it can result in population decline on a scale observable in some parts of the developing and industrialized world where institutions deteriorate because the people necessary to support them are no longer present. It is clearly inadequate to talk simplistically of 'push' and 'pull' factors or even to seek local causes of emigration, which is a general phenomenon. Arizpe suggests that it is more useful to analyse the departures from the general trend than to attempt more detailed analysis of the process of migration.

A general statement concerning the underlying causes of rural emigration does arise as a consequence of these three essays. The pervasive acceptance by rural populations of the superiority of urban life, whatever their social category and, often, whatever their ethnic origin, is a consequence of the penetration of rural homes by modern communications media, especially the radio. These media are both urban-based and city-biased, and the messages are designed to stimulate demand for the products of major industrial organizations, controlled from the industrialized countries, whether they be Pepsi-Cola, Philips radios, or a Bayer insecticide. Equally potent and pervasive purveyors of non-local goods are teachers, priests, and other outsiders living in rural communities. Both consciously and unconsciously they display their knowledge and appreciation of such products and their ignorance of and suspicion for local products. With increasing contact with such people, in positions of power and authority, local people imitate their tastes and acquire goods that their parents neither wanted nor perhaps needed. The necessary purchasing power can only be obtained by producing more or from working elsewhere for a wage, and thus migration becomes more necessary and desirable.

Tangential to this statement, we must recognize that the production of many goods which are used by rural families is taking place on a large scale at a lower cost than was previously possible, thus lowering prices and making unprofitable other industries employing simpler technology but, in total, many more people. Such industries were often located in rural areas. The production of some major export crops is largely the responsibility of larger-scale farmers and to a varying extent controlled by exporting agencies of multinational firms. Thus, even where small-scale farmers produced a crop such as coffee primarily for export, the demands of the external market require the farmer to produce a product of a specific quality, and individuals may be hard put to do this unless they group together as a cooperative or sell their product to a processor at a disadvantageous price. A set of factors brings rural families into closer contact with economic and cultural forces over which they have little control and which combine to encourage change or withdrawal from farming and the rural world.

Finally, we may note the inability of national Governments to encourage the development of rural areas in such a way that more, rather than fewer, opportunities are created there. Recognition of the implications for national prosperity of the migration of the rural population is only partial, and few clearly formulated policies exist which seek to offer alternatives to migration. This is, in part, because national governments represent urban and, especially, metropolitan interests and the development of the countryside inevitably means less growth for the urban centres where almost all Latin American Governments have their power base.

EXTERNAL PENETRATION

A preoccupation of increasing importance to students of rural affairs in the developed and underdeveloped world is the increasing control over farming that is exercised by large companies. In the United Kingdom, this is brought about by the acquisition of top-quality farmland by finance companies and by the amalgamation of farms and their control by farming organizations concerned with short-term profit maximization. These changes have been barely studied by social scientists, but similar changes in North America and Latin America have attracted more attention, and the probability of such changes increasing in importance in the future merits attention in this volume.

The perception by Latin Americans of external penetration is variable. On the one hand, foreign influences are evident in manufactured goods which are widely publicized and whose acquisition is commonly a symbol of status. In the declining industrialized countries, accustomed to purchasing many manufactured goods produced nationally, the increasing proportion of such goods imported or produced nationally by factories controlled by overseas interests may pass unnoticed. In less-developed countries, however, the influence of foreign business interests is widespread, and its spread is in part associated with a larger proportion of the population becoming a part of the market for manufactured goods as a result of migration to major cities, of the decline of cottage industries, and as a consequence of small-scale agriculture becoming more commercially oriented and family self-sufficiency being less prized and also less possible. The rural population is becoming increasingly part of the consumer society and, spurred on by advertizing and the images that are presented for emulation in national media, many rural people wish to be able to become commercial farmers, or, more likely, workers for commercial farmers earning wages that may allow the acquisition of the consumer goods that most represent social progress and which will embellish the self-image of the members of the family.

Feder's article emphasizes the most recent trends in foreign involvement with rural growth and describes it as the relocation of United States agriculture in Latin America. This parallels the relocation of some United States manufacturing industries in Mexico, particularly along the United States–Mexico border. This analysis demonstrates the complexity both of the responsibility for management of production and also of the means by which 'aid' from developed countries particularly benefits commercial interests in the aid-giving country. Even within the recipient country, beneficiaries are often more diverse than those for whom the aid was intended. A further important part of Feder's analysis is the demonstration that world-scale price movements seldom harm corporations involved in handling major commodities. This is currently demonstrated in the industrialized world by the increasingly large profits made by petroleum companies in the face of price rises by exporting countries, even though many countries have a considerable measure of nominal control over petroleum processing and retail sale through part-ownership of major oil companies and control of taxes.

Two points, which are seldom considered in the literature on the subject, are worthy of examination in the context of external penetration. In the first place, it is not easy to see how people in Latin America can materially improve the quality of their life within the capitalist system without participating in the sectors of the economy that are dominated by external interests. Even if external interests exclude purely national sources of capital, many of the goods most freely available for consumption and much of the technology that is most accessible for use is dependent on overseas financial and commercial interests. Social development in the provision of schooling, medical services, and communications is completely involved with value systems and technology developed outside Latin America. Non-formal education, greatly publicized through the work and teaching of Paulo Freire in Brazil, is often spread through the work of foreign-financed organizations—in Ecuador, by the University of Massachusetts, for example.

Material improvements in personal and community well-being can be achieved without furthering external penetration only by the development of national and regional human resources and by the mobilization of popular interest to encourage the search for ways towards self-improvement which use national resources without losing any control over them. The Catholic Church has so far shown most dynamism in the development by bishops and clergy of means for encouraging the articulation of popular needs and helping to help people find the most satisfactory means to satisfy them. At a national level, little has been achieved to strengthen economies and to reduce dependence on outside assistance. If there are few alternatives to external aid for social and economic development that have been demonstrated, continuing foreign domination is as likely as Feder suggests.

In the second place, and in various ways related to the previous issue, it is necessary to determine the extent to which external penetration is actively sought by developing countries and whether there are *any* ways in which outside assistance can be obtained which do not involve the abandonment of national control.

A very small proportion of foreign commercial contact is the result of altruistic motives. British foreign policy is consciously concerned with promoting British interests, and recent political disagreement concerning the freedom of the Overseas Development Ministry from control by the Foreign and Commonwealth Office was concerned largely with the latter's preoccupation with ensuring that Britain derived as much benefit as possible from its aid programme. Even though profit can be made by expansion of an industrialized nation's commercial interests to Latin America, the terms offered by different firms are equally disadvantageous. Even where agreements include the setting up of industries in a specific country, the initial investment and profit are recovered in a very short time and the industrial plant and its products are not necessarily those most suitable for the needs of the country in which it has been established. There may, therefore, be few ways by which Latin American countries can obtain assistance and investment without putting themselves at a disadvantage.

External penetration is actively sought both by foreign interests and by Latin American Governments. The better-off and more-articulate sectors of the population demand goods and services which can most easily be provided from overseas, and available local capital is either invested directly overseas or invested to use imported technology in Latin America. For their part, foreign interests are concerned primarily with maximizing profits, and Latin America has long been an area where profits are made quickly, even if instability makes longer-term profitability of foreign investment less certain.

In conclusion, one may seriously doubt whether a viable alternative to the present ways of development can be developed by the industrialized capitalist world, nor even by the Soviet Union, which has long encouraged dependent development of regions and States associated with it. In particular, the sort of technology currently being used in the West may not be suitable for use in Latin America, just as many doubt its suitability even in Western Europe. National Governments need to concentrate on an identification of solutions to major problems, and this involves both the range of known national resources and the national population, and at a regional as well as a national level. If external trade is reduced, then its effects on the domestic economy can be minimized and controlled.

In our consideration of rural futures in Latin America, it is worthwhile seeking to identify alternative strategies for the future which hold more hope than contemporary policies can offer.

NOTE

[1] These studies were carried out for a consortium of international agencies (OAS, FAO, IBD, ECLA, IICA) by the Inter-American Committee for Agricultural Development (CIDA) who published the case studies, and the results were analysed in Solon Barraclough (Ed.), 1973, *Agrarian Structure in Latin America* (Lexington, Man.: D. C. Heath). Results of the second-stage studies were diffused informally except for the massive report of the Mexican study contained in Sergio Reyes et al., 1974, *Estructura Agraria y Desarrollo Agricola en Mexico* (Mexico: Fondo de Cultura Economica), and no final reports were ever published for the Venezuela or Bolivia studies. The Land Tenure Center Library (Madison, Wisconsin, U.S.A.) does have a collection of internal papers relating to these projects.

BIBLIOGRAPHY

Browning, D., 1971, *El Salvador: Landscape and Society* (Oxford: Oxford University Press).

CIDA, 1964, *Ecuador: Tenencia de la Tierra y Desarrollo Socio-Económico del Sector Agrícola* (Washington: Unión Panamericana).

Collier, G. A., 1975, *Fields of the Tzotzil. The Ecological Bases of Tradition* (Austin: University of Texas Press).

Dozier, C. L., 1969, *Land Development and Colonization in Latin America* (New York: Praeger).

Favre, H., 1977, 'The dynamics of indian peasant society and migration to coastal plantation of coastal Peru'; in Duncan, K., and Rutledge, I. (Eds), *Land and Labour in Latin America* (Cambridge: Cambridge University Press) pp. 253–67.

Kleinpenning, J. M. G., 1977, 'An evaluation of the Brazilian policy for the integration of the Amazon region 1964–74', *Tijdschrift voor Economische en Sociale Geografie*, **68**, 297–311.

Preston, D. A., 1974, 'Geographers among the peasants: research on rural societies in Latin America'; in Board, C., et al. (Eds), *Progress in Geography*, No. 6 (London: Arnold) pp. 143–78.

Roberts, B., 1978, *Cities of Peasants* (London: Arnold).

Roberts, R. L., 1975, 'Migration and colonization in Colombian Amazonia: agrarian reform and neo-latifundismo', *PhD Dissertation*, Syracuse University.

Salomon, F. L., 1978, 'Ethnic lords of Quito in the age of the Incas: the political economy of the north-Andean chiefdoms', *PhD Dissertation, Cornell University*.

Stewart, N. R., Belote, J., and Belote, L., 1976, 'Transhumance in the central Andes', *Annals of the Association of American Geographers*, **66**, 377–97.

Land Reform

Environment, Society, and Rural Change in Latin America
Edited by D. A. Preston
© 1980 John Wiley & Sons Ltd.

CHAPTER 2

Land reform, livelihood, and power in rural Mexico

Cynthia Hewitt de Alcántara

INTRODUCTION

Agrarian reform is a crucial element in any strategy that pretends to raise levels of living in rural areas where land is so inequitably distributed that the livelihood of the majority is prejudiced by the privileged position of a minority. This is the case not only because access to some vital minimum of physical resources must be guaranteed to the majority but also because the power which accrues to large landowners by virtue of their oligopolistic position must be distributed more evenly among the population at large if the participatory dimensions of well-being are to be reinforced. Agrarian reform implies, in other words, not only a redistribution of land but also a redistribution of power.

The latter dimension of the process is extremely complex, and lends itself all too often to forms of manipulation which lessen the real benefits obtained by rural people from the redistribution of land. At a most basic level, of course, the delivery of a plot of land to a family previously required to work for someone else, or on his conditions, must represent an increase in the capacity of that family to make decisions that can positively affect its livelihood. There can be no doubt that it is very much better to receive land than not to receive it. However, to the extent that each new land reform beneficiary is integrated into a wider structure of economic and political power, he must also be able to count upon a more generalized position of bargaining strength in order to ensure that the terms of exchange for goods and services provided by the latter are not systematically unfavourable to him. Without such protection, implying political organization, he may eventually find himself exploited indirectly by any number of politically and economically more powerful elements within the wider society, much as he was exploited directly before reform by large landowners.

Paradoxically, the very process of agrarian reform contains within it mechanisms that can, under some conditions, reinforce the exploitative capability of the wider society, and which must be consciously counteracted for the positive potential of reform to be fully realized. The roots of the problem lie in an inevitable increase in the authority of the State in rural areas following land redistribution. To understand this development, one must remember that land reform is more than the transfer of property from one group (a minority) to another (a majority). Such a transfer can take place through the direct appropriation of a contested holding by the peasantry; but the action is legitimized only through the intervention of non-peasant groups in control of the apparatus of the State. This is, in fact, the fundamental difference between a 'land invasion' and a 'land reform'. A land reform is a redistribution of property sanctioned by the authority of the State. And, in the process of conferring legitimacy upon a new agrarian order, the State gathers for itself allegiances formerly owed by land reform beneficiaries to large landholding interests or, more significantly, to peasant organizations long engaged in struggling against large landowners. The State, in other words, is put in a position to exert maximum power in the countryside, to challenge both the authority of the original agrarian élite and that of the organized peasantry.

The way this power is utilized by groups within the Government and the bureaucracy depends upon the ideology of dominant factions within the State, upon the balance of power within the wider (national) society, and upon the constraints to action imposed by the position of the nation within the international community. Taking advantage of the superordinate position guaranteed them in the countryside by a monopoly over the legitimization of agrarian change, representatives of the State may support an ongoing process of peasant organization, integrated into a broader programme designed to better the livelihood prospects of low-income groups in rural areas. Or they may divide and disorient the peasantry, in an effort to prevent the voicing of rural demands, while pursuing policies primarily intended to increase capital accumulation within urban, industrial sectors. If the first course is followed, agrarian reform fundamentally contributes to ensuring equality of opportunity for all rural people; and, in fact, the peasantry as a low-status group tends to disappear. If the second course prevails, however, the redistribution of land will not be accompanied by an effective increase in the power of the peasantry, which is likely to continue even after agrarian reform as a low-status group living in a disadvantaged relation to the national society and economy. In this latter case, the re-initiation of a process of dispossession and proletarianization is virtually inevitable, as is the renewed voicing of peasant protest.

Let us see how these generalizations apply to the case of Mexico.

THE FIRST PHASE OF AGRARIAN REFORM IN MEXICO: A LIMITED RECONSTRUCTION OF THE PEASANTRY

Agrarian reform began in Mexico at the insistence of a peasantry in arms, angered by the unremitting encroachment of neighbouring large landowners on communal lands and made desperate by the rapid decline in levels of living which occurred in areas of greatest agricultural modernization towards the end of the nineteenth century. Under the leadership of Emiliano Zapata, villagers living in the densely settled central region of the country went to war in 1911 for the principles of 'land and liberty' (*tierra y libertad*); and they were soon followed by members of countless other rural communities throughout Mexico. 'Liberty,' in this context, did not mean parliamentary democracy. It meant autonomy, the right to conduct community affairs without undue outside interference, the possibility to reinforce the foundations of a peasant order badly shaken by the incursions of an advancing capitalist society. The return of communal lands was the single most important element in ensuring this autonomy. Therefore, as early as 1915, the Zapatistas began to supervise the formal reoccupation of conquered territory by members of communities to which it had originally belonged.

Such a redistribution of property, accomplished with very little paperwork on the basis of personal consultation among representatives of communities with traditional rights to the land in question, epitomized a peasant ideal of agrarian reform. It was not destined to endure, however, for the peasantry of twentieth-century Mexico could not ultimately dispose of a portion of the national territory without the assent of the State. Until legitimized by the new liberal élite which came to power in the aftermath of the Mexican Revolution, the reoccupation of communal property by the peasantry was considered illegal. It was ruthlessly undone between 1916 and 1919 by federal troops under the command of non-peasants who proclaimed their own 'agrarian laws' and applied them to the conquered countryside as if no previous effort had ever been undertaken. In order to receive title to lands wrenched from their control by the *haciendas* before the Revolution, peasant communities were forced to become the petitioners of the State.

The latter, particularly during the chaotic period immediately following the end of the Revolution in 1917, was far from monolithic. At the national level, a number of personalistic factions struggled for control of the administrative machinery which would allow them to impose their mandate upon the country; and, at many subordinate levels reaching downward towards the countryside, shifting alliances were formed to bolster the power of one faction or another. In such a situation, the peasantry constituted a vital political resource. All major groups contending for national power therefore promised to return communal lands to their traditional owners, less out of conviction, in some cases, than out of political necessity. Land was exchanged for allegiance, which was translated into power.

During this first phase of agrarian reform in Mexico, none of the groups which successively gained control at the national level supposed that lands returned to peasant communities would be utilized to contribute significantly to national agricultural production. On this point, their view coincided with that of the majority of the peasantry: returned plots were vitally needed to contribute to local self-provisioning and were not dedicated to commercial crops on any scale worthy of note. The post-Revolutionary élite did not find such a development immediately disturbing, because it was assumed that commercial agricultural production would remain the province of large holdings, most of which had not been affected by agrarian reform. This assumption is clearly reflected in statistics from the agricultural census of 1930: during the fourteen years following the first official grant in 1916, some 4000 communities received land, but their holdings counted for only 6 per cent of all the farmland of the country. Remaining land belonged to private owners, 56 per cent of whom held 10 000 hectares or more each.

This was, then, a period of temporary respite for villagers in communities which were able successfully to press their claims for the return of usurped lands, and thus to engage once more in subsistence farming with an increased possibility of meeting their basic needs. It represented a limited reconstruction of the peasantry in certain regions, permitted in return for the demobilization of peasant militias and submission to the authority of new groups within the élite of the nation.

But the majority of the peasant population of Mexico, living in less militarily or politically strategic regions of the country, or integrating the rural proletariat of large capitalist holdings, was not included in these early efforts at agrarian reform. The allegiance of many rural people had therefore not been captured by the post-revolutionary national State. And, as levels of living in the countryside declined under the impact of the world depression (particularly disastrous in its effect upon inhabitants of centres of export agriculture), peasant unrest threatened the shaky peace upon which that State was based. At the same time, large landowners, uncertain of their future, failed to produce sufficient goods to satisfy urban demand, or to check the rapid advance of inflation. A comprehensive reorganization, both of agricultural production and of peasant political participation, was urgently needed; and that required a new conception of agrarian reform.

THE SECOND PHASE: AGRARIAN REFORM AS INTEGRATED RURAL DEVELOPMENT

During the 1930s, the idea that the peasantry should participate only marginally in the economic and political system of the nation was exchanged for the conviction that it should be fully integrated into national life, as producer of commercial agricultural products, consumer of manufactured goods, and citizen. This radical change in the position of the peasantry could be wrought, however, only through the massive redistribution of land still held in large properties, the formation of a modern structure of organizational support for small farmers, and the integration of all small cultivators into a political structure capable of making their influence felt at the regional and national levels. Within such a framework, agrarian reform took on the characteristics of an integrated programme of rural development and implied the investment by the State of large sums of money in the countryside.

Between 1934 and 1940, during the presidential administration of Lázaro Cárdenas, more land was distributed in Mexico than at any other time since the revolution. A substantial part of it was expropriated from the richest commercial farms of the country and delivered to agricultural labourers organized in collectively owned and operated agricultural enterprises known as 'collective *ejidos*'. These enterprises were advised by employees of a newly founded Ejido Credit Bank, which financed highly remunerative commercial crops and provided the technical assistance necessary to recover its investment. In the last analysis, it was the Bank which replaced the original private owners of expropriated properties as entrepreneur, and which partially substituted the rôle left vacant by hired managers on the large estates as well. The potentially dominant position of the State, through the Bank, was temporarily held in check, however, by the consolidation of strong regional peasant organizations and the creation of a National Peasant Confederation which effectively represented peasant interests for a short time during and immediately following the Cárdenas years.

A number of accounts of this brief experiment in the collective organization of land reform beneficiaries in the best-endowed regions of the Mexican countryside document remarkable improvement in levels of living during the late 1930s and early 1940s, as well as increasingly effective local participation in making the decisions on which livelihood depended. At the same time, the collective *ejidos* managed, on the whole, to meet and surpass the level of production set by private estates before expropriation, and thus to prevent the disruption of the agrarian economy so emphatically predicted by opponents of agrarian reform. Regional peasant organizations in the most famous centres of collectivism, La Laguna and the Yaqui Valley, in fact grew so powerful that they began to enter the field of complementary industries and services, owning and operating cotton gins, electric plants, urban bus lines, and insurance companies.

Members of collective *ejidos* did not, however, by any means constitute the majority of all peasants receiving land during the Cárdenas administration. The former were a privileged and highly visible minority on whom most of the budget available for rural development was spent in an attempt to prove that agrarian reform and increasing agricultural productivity could be compatible. A second, and much larger, group of beneficiaries was formed by other landless labourers or dependent cultivators associated with smaller and more isolated estates who received expropriated land in the form of non-collective *ejidos*.

The particular form of agrarian organization known as the *ejido* was to some extent an outgrowth of traditions of communal landholding which could be traced as far back in history as the Aztec *calpulli* and the Spanish colonial *ejido*, or free pasture and woodland, from which the name was drawn. But it was the Zapatistas who gave to the word *ejido* the more general meaning of recaptured communal lands of any kind (including those destined to cultivation), and the national agrarian legislation of the 1920s which made the *ejido* the principal institution through which the State could grant land to petitioning agrarian communities. Land so granted became the corporate patrimony of each community, to be worked either individually or collectively, as the community preferred, and (according to the Agrarian Code of 1934) to be administered by a six-man committee elected every three years. Most *ejido* land was, in fact, worked in individual parcels by beneficiaries who passed their holdings on to an heir at their death, but who were nevertheless forbidden by law to mortgage, rent, or sell their land, on pain of forefeiture to a new beneficiary designated by the *ejido* community as a whole.

Such an arrangement made the *ejido* something more than a type of land tenure; it became an instrument of local government as well, a new basis for cooperation along modern lines, involving patterns of authority quite different from those found within traditional rural communities. The ceremonial *cargo* (or office) system of traditional communities assigned prestige on the basis of length and degree of devotion to public service, and therefore gave deference to age. It was a participatory system, but not one based on parliamentary rules. The *ejido*, on the other hand, was governed by a group of officials elected in an open meeting by majority vote. The purpose of this innovation was to school rural people in the practice of parliamentary democracy. And, most particularly in regions where non-indigenous land reform beneficiaries were grouped into entirely new communities in order to form an *ejido*, there can be no doubt that such a goal was significantly furthered. But the more remote and indigenous the community, the more likely it was that the introduction of a new structure of authority which competed with pre-established patterns would foment division and conflict, at least as much as local democracy. This was particularly the case because *ejido* authorities had to be adept at maintaining contact with representatives of the national Agrarian Department, ultimately charged with ensuring that *ejido* lands were utilized according to the provisions of the law. Such a requirement gave power to the most educated and affluent members of a community, and to those most successful at building political alliances with the outside world.

The provision that *ejido* lands could not be mortgaged, rented, or sold reflected the long-standing determination of all those concerned with improving peasant livelihood, from the time of Zapata onwards, that the gains of agrarian reform not be jeopardized by subjecting the land of peasant beneficiaries to the whims of the market. By granting title to *ejido* land to the community, and placing its ultimate disposition under the protection of the State, a reconcentration of holdings was to be avoided and the subsistence requirements of peasant families assured. This has since been repeatedly criticized as introducing an element of rigidity so formidable that the efficient allocation of resources in the countryside has been impossible. Such an argument, however, lays the blame for subsequent difficulties within *ejido* agriculture at the wrong door: the kind of agricultural development envisioned by Cárdenas, and others of his persuasion, was entirely possible within a framework of *ejido* landholding. The critical elements for its success were effective local organization and the provision of adequate financial and technical assistance by the State. When later political changes at the national level impeded both organization and access to needed resources, the productive potential of most non-collective *ejidos* was not realized. Nevertheless, the level of living of peasant families in many parts of the country rose markedly with the receipt of *ejido* land. Widespread subsistence farming on inalienable plots was certainly preferable in social terms to the wave of dispossession which would have followed the derogation of *ejido* law and a return to the unrestrained operation of a free market in land.

The Cárdenas administration came to a close in 1940 under very difficult economic and political conditions. The nationalization of the petroleum industry in 1938, the massive expropriation of private agricultural property (often in foreign hands), as well as the more general orientation of the government toward the encouragement of workers' and peasants' organization, and its manifest sympathy for international socialism, won it the decided distrust of both foreign and domestic investors. Foreign investment declined precipitously as the Cárdenas period went on, and domestic capital fled the country. In the resulting crisis, the economic resources at the disposal of the government were not sufficient to continue financing a far-reaching rural development programme. And the political resources on which the administration had to count in order to forestall an impending challenge from the right grew progressively smaller under the impact of economic recession. Urban lower-middle and middle income groups, in particular, were disaffected by official support to peasants and workers, and threatened to cast their lot with the right. Short of re-arming the peasantry and workers, and rekindling the revolution, Cárdenas had little choice but to sanction the assumption or power in 1940 by a group which could win back the confidence of private capital and ensure the kind of economic growth desired by the urban middle and upper classes. The end of his presidency, therefore, coincided with the abandonment of the strategy of integrated rural development for which it had stood, although agrarian reform as a piecemeal programme of land distribution was to continue for decades to come.

THE THIRD PHASE: AGRARIAN REFORM AS AN INSTRUMENT OF POLITICAL MANIPULATION

The official abandonment of the Cárdenas strategy of development after 1940 implied a discontinuation of some rural development programmes, a drastic reduction in the budgets of others, the systematic undermining of strong peasant interest groups, and eventually the use of force against centres of independent peasant political power. It also marked the beginning of renewed support for large private holdings, once again given a key rôle in supplying the agricultural products required by an industrializing urban society. A corporate State, firmly controlled by the private sector, was in the making; and, in the policy of such a

State, there was no place for a peasantry capable of asserting independent economic or political influence.

It was nevertheless impossible for governments following Cárdenas to ignore entirely the legacy of his administration. Unlike the situation in 1930, when land reform beneficiaries still stood at the margin of national life, the *ejido* in 1940 was a permanent and central aspect of socio-economic organization of the Mexican countryside. Not only did millions of members of peasant families live in communities governed internally by *ejido* authorities and tied through administrative networks to national banks and ministries charged with overseeing *ejido* affairs, but slightly over one-half of all the agricultural production of the country was attributable to the effort of *ejidatarios*, including an important group of land reform beneficiaries collectively producing some of the most remunerative export crops of the nation. No matter how strong the post-Cárdenas preference for private enterprise and how great the distrust of the peasantry, public policy was forced to contend with a *fait accompli*: *ejido* land could not be taken away from peasant beneficiaries, whose gratitude to President Cárdenas could easily turn toward a militant defence of his programme.

Agrarian reform, as a political and administrative instrument, was therefore not abandoned in the decades to come. On the contrary, post-Cárdenas governments systematically utilized the institutional infrastructure inherited from the Cárdenas period to increase the economic and political control of the State over the countryside. Organizations associated with land reform which originally had possessed some capacity for independent local initiative, some degree of control over local resources, were reoriented to ensure that they would meet first the needs of an urban industrial order. In the best-endowed regions of export agriculture, where collective *ejidos* had been created during the late 1930s, for example, regional peasant interest groups not controlled by the single official party were destroyed, their members denied access to credit, and their leaders at times assassinated. The collective farms were parcelled out among individual *ejidatarios*, with neither the organizational nor the financial capacity to work them alone. And then, in order to keep these economically strategic centres of agrarian reform in production, peasant plots were grouped into small credit societies and virtually farmed by the Ejido Bank. The latter determined dates of planting, irrigating, and applying manufactured inputs; it supplied those inputs without seriously consulting *ejidatarios* and charged the cost to the accounts of its clients; it had the sole authority to receive and sell the output of its clients' fields, and to deduct often unexplained amounts before distributing similarly unexplained 'profits'. In a very real sense, the State became a new patron; and land reform beneficiaries of irrigated regions found themselves in an uncertain position of dependence, obviously higher in status and in income than that of peon, but unable to make the most basic decisions concerning the management of their land.

Throughout other regions of the country, where collective *ejidos* had never been established, but where local resources were nevertheless of sufficient importance to the national economy to require attention, the Ejido Bank played a rôle similar to that just described for the principal irrigation districts of Mexico. Credit was supplied on a short-term basis in order to ensure the production of required commercial crops; and the crops were channelled out of the countryside toward urban consumers, urban industry, and export warehouses. Local needs were given little priority. Credit for medium- and long-term investment in *ejido* agriculture was virtually absent; credit for the establishment of small rural industries oriented toward supplying local markets entirely so.

As time went on, the political importance of short-term credit became increasingly clear. Not only could this vital resource be utilized to strengthen leaders and groups associated with the official party, but it could also serve as a more general instrument in buying rural tranquility and preventing outbreaks of protest impelled by declining levels of living. The henequen region of Yucatan, for example, suffered the effects of a prolonged economic depression after World War II. Instead of undertaking the diversification of the one-crop economy—a step opposed by the oligarchy of Yucatan, known as the Divine Caste—the Ejido Bank was charged with making weekly payments to tens of thousands of land reform beneficiaries so that they could continue to work in henequen. Credit to henequen *ejidos* became, in effect, an income subsidy; it prevented starvation, though not extreme deprivation, and it was increased in times of most violent peasant protest against the henequen system. Similar forms of subsistence (or survival) payments were made in the form of credit to scattered groups of *ejidatarios* in the northern desert, eking out their living by gathering cactus wax and fibres.

The primacy of political considerations in the administration of credit implied a strong tendency towards inducing dependence, even at the cost of economic efficiency. Short-term credit to perfectly credit-worthy *ejido* enterprises was treated as much as a dole as the true income subsidies just described. And since effective participation by *ejidatarios* in the management of well endowed commercial plots was so systematically discouraged, their receipt of weekly credit advances did, in fact, tend to take on the appearance of a gift from the State. This trend was furthered by the noteworthy inefficiency, and not infrequent dishonesty, with which the Ejido Bank, as well as other State agencies servicing the land reform sector, managed *ejido* resources. The technical level of personnel was low, and salaries still lower. Therefore, the temptation to receive commissions for buying inputs at slightly higher prices than that which might be obtained competitively, for investing *ejido* funds in private short-term ventures between crops, or for selling *ejido* crops at disadvantageous prices was very great. Even in the wealthiest land reform areas, most *ejidos* were soon heavily in debt, and their members accustomed to the status of debtors.

Such a systematic misuse of the human and physical resources of the countryside, unintelligible in purely economic terms, was a clear reflection of the central dilemma of post-Cárdenas governments. In order to ensure the kind of urban industrial growth then being promoted within a wider framework of dependent capitalism, agricultural goods were needed not only from large private holdings, but from well endowed land reform holdings as well. Yet any effectively independent development of the latter would jeopardize centralized political control of the countryside, allow the voicing of protest against the urban bias inherent in post-war modernization, and perhaps threaten the political peace which made that modernization possible. In such a situation, the short-term extraction of agricultural products from a disorganized and dependent peasantry was favoured over long-term investment in viable peasant enterprises.

The total number of land reform beneficiaries with whom the Ejido Bank worked in any given year between 1950 and 1970 averaged only 32 000, or roughly 2 per cent of all the *ejidatarios* in the country. The vast majority of the small cultivators of Mexico, including a significant group of peasants working privately owned plots (some 610 000 holdings in 1970, often bought during the same process of breaking up large landholdings which had produced the *ejidos*), therefore had to depend almost entirely upon local moneylenders for financing any part of their agricultural operations which could not be covered with the resources of the family alone. Some, with sufficient productive potential to interest private companies dealing in manufactured inputs, or wholesale buyers of agricultural commodities, became the clients of these intermediaries, entering into contracts which promised the delivery of crops at a low price in return for credit. The rest formed the mass of increasingly impoverished *minifundistas*, in possession of holdings created by the political necessity to distribute land but entirely excluded from any programme which might have provided the organizational, technological, and physical resources needed to defend a certain minimum level of family income.

During the period between the end of the Cárdenas administration and 1970, land was granted to an average of 210 000 petitioners during each six-year presidential term. The manipulative element inherent in this continuing distribution of land is illustrated not only by the marked lack of official attention to making most smallholdings productive, but also by the frequency with which grants were only partially legalized, leaving a wide margin for manoeuvre on the part of representatives of the official party and State-run agencies interested in eliciting loyal behaviour from the peasantry. The legal process through which petitioners for land were required to pass by successive agrarian codes, in order to obtain definitive title as *ejidatarios*, was so complicated that neither the petitioners themselves nor the agrarian bureaucracy had the resources to ensure full compliance with the law. Thousands of cases went unattended, or were shuffled from one agency to another, as land reform beneficiaries waited to receive some guarantee of their rights. According to one recent study, the average amount of time which has elapsed in Mexico between the original petition for land and receipt of definitive title is nine years, plus an additional five years before the boundaries of individual *ejido* plots can be fixed. And, during the early 1970s, approximately 40 per cent of all the grants of land made since 1916 were still not legally completed.

Such an arrangement, like the administration of official credit discussed above, promoted a politically advantageous dependence upon the State at the cost of economic and social rationality. Not infrequently, the amount of money required to continue pressing individual and communal claims in the state capital, and ultimately in the Federal District, represented such a considerable part of all available monetary resources of poorer communities that little was left over for more productive purposes. In addition, *ejidatarios* without legal title were prey for neighbouring large landowners and enterprising local *caciques* intent upon expanding their holdings. And a number of *ejidos* were plagued by internal violence, springing from misunderstandings concerning the real boundaries of individual plots. Until forced by desperation into acts of protest against the inefficiency of the agrarian bureaucracy, however, most *ejidatarios* attempted to maintain good relations with representatives of the official party and the State. The latter represented the only hope that a definitive resolution might some day be forthcoming.

DETERIORATING TERMS OF LIVELIHOOD IN THE POST-WAR PERIOD

The legal insecurity clouding livelihood prospects in so many agrarian reform communities was only one element of a more general, structural insecurity which increasingly characterized social relations in the Mexican countryside during the post-war period, and which made it less likely in 1965 than in 1935 that the receipt of a small plot of land would suffice to provide for the subsistence needs of a rural family. The roots of growing insecurity were to be found in the collapse of barriers (physical and mental, economic and political) which had served to prevent the full integration of peasant communities into the national economy and society, and thus had preserved a degree of local control over the terms of livelihood.

For centuries, rural people throughout Mexico had survived by emphasizing micro-regional self-sufficiency in basic goods, limiting trade with the wider monetary economy, curbing conspicuous consumption within the village, and enforcing a limited redistribution of small surpluses through the periodic sponsorship of religious celebrations. The resources of most communities were not great, but neither were their needs; and a web of personal relationships among local families functioned to ensure that no-one would be destitute (unless a catastrophe left the entire village destitute).

With the communications revolution of the post-war period, when even the remotest rural communities were reached by motor roads, buses, and radios (and consequently by politicians, middlemen, and schoolteachers), the balance between local resources and needs was quantitatively and qualitatively altered, at the same time that traditional mechanisms of mutual assistance disintegrated. Rural people began to see themselves increasingly through the eyes of a modern urban culture, and the picture was not at all complimentary. To escape ridicule, and perhaps to win respect, within the wider society, it was necessary to compete for status on urban terms: to have money and to spend it on—more nearly citified housing, clothing, and durable consumers goods. Yet money was not at all easy to obtain. It required selling local goods or labour in a wider market, at prices determined by the interplay of factors over which no local control could be exercised. As a result, the increasing use of money tended to imply a widening field for exploitation: the value of goods and services provided by rural families to a modern urban society was likely to be consistently greater than the value of goods and services returned.

Concurrently, the need for money came to characterize not only exchange relations between countryside and city, but those among rural families as well. This constituted a serious challenge to subsistence in many areas, where small plots could provide for the needs of a family only if extra hands required at peak seasons could be obtained through the cooperative exchange of labour among neighbours. Mechanisms of unpaid exchange did not disappear entirely in all peasant communities; but the increasing demand for wage labour on commercial farms and in the city, made effective in formerly isolated areas with the coming of paved roads and buses, forced local cultivators to pay more and more frequently for field help with money. Even the most exiguous expenditures for this kind of assistance, often combined with other new monetary costs of production (for the rent of a yoke of oxen, for example, formerly exchanged among kinsmen without payment), sufficed to eliminate precarious margins of production above subsistence and to plunge peasant families into debt. Debt, in turn, implied high rates of interest paid to local moneylenders, thus compounding debt. Peasant families caught in this web of dependence had little choice but to sell their corn, wheat, or beans at a low price immediately after harvest, in order to meet the monetary costs of production, and then to buy it back shortly thereafter at a higher price imposed by speculators, in order to eat. And if they attempted to break out of the circle by planting apparently more remunerative commercial crops like coffee, tobacco, or vegetables, the higher costs of inputs, and therefore greater dependence on credit, often served to compound livelihood problems.

The situation was made worse by population growth (reaching a national rate of 3.4 per cent per year during the post-war period), and by the progressive elimination of many non-agricultural sources of income in the countryside. Small rural industries were liquidated by competition from manufactured goods: pottery gave way to plastic and enamel, homespun cloth to synthetics, *pulque* to beer. At the same time, a number of traditional service occupations were made irrelevant by modernization, among them muleteering and various forms of commerce. The boundaries of gainful employment shifted rapidly to the cities, where prerequisites of education and urban *savoir-faire* placed them often beyond the reach of rural people.

In order to survive, most of the smallholders of Mexico (including the majority of all land reform beneficiaries) became a new kind of semi-proletariat, working in the cities or on large commercial farms for a sufficient number of days to obtain the money income required to sustain the yearly planting of a subsistence, or infrasubsistence, plot. That plot was in most cases not abandoned, for it constituted the difference between survival and starvation; the wider economy offered no promise of employment sufficiently remunerative to support a peasant family on a permanent basis. But neither could the plot be capitalized. Agricultural census figures illustrate the magnitude of the dilemma: in 1960, as in 1950, 85 per cent of all holdings in Mexico were classified as either infrasubsistence or subfamily, producing together only 28 per cent of the farm output of the country; and figures contained in the 1970 census suggest that the proportion of holdings in these two categories changed little, if at all, during the following decade.

THE FOURTH PHASE: AN ABORTIVE RETURN TO INTEGRATED RURAL DEVELOPMENT

Not surprisingly, then, in the 1970s, as in the 1930s, political and economic crisis in the countryside once again forced a re-evaluation of agrarian policy at the national level. The post-war strategy of relying heavily upon the production of large private farms, while simultaneously extracting commercial agricultural products from a disorganized and disillusioned remnant of the better-endowed *ejidos* created during the Cárdenas period, and abandoning all other land reform beneficiaries to survive as best they could, had been successful in maintaining urban industrial growth and rural peace for almost three decades. But the arrangement contained logical limits, which were eventually reached. By the early 1970s, the annual rate of growth of the agricultural product, averaging 4.6 per cent between 1942 and 1964, dropped to 0.2 per cent. Declining returns on capital investment in private commercial agriculture played an important part in this change, as did the steadily worsening position of smallholdings. At the same time, the ecological and financial price of dealing with waves of migration from countryside to city became clearly prohibitive. The political implications of increasingly violent confrontations between a privileged minority and an impoverished majority in rural areas could no longer be ignored. A return to the Cárdenas strategy of integrated rural development seemed imperative.

Such a change in policy proved impossible, however, because rural development as promoted by Cárdenas implied redistribution, both of wealth and of land; and in the 1970s, structural impediments to this form of attack on rural poverty in Mexico were at least as great—if not greater—than in 1934.

The best land of the nation was rented or owned by a powerful group of entrepreneurs, so thoroughly integrated into the commercial and banking élite of the country that any attempt to expropriate their holdings (at times surpassing the legal limit by thousands of irrigated hectares), or to challenge their position as commercial intermediaries between countryside and city, was likely to bring on a national economic and political crisis. Crisis was also implicit in any effort to promote the kind of local political organization of the peasantry required to increase the productivity, and to defend the bargaining power, of those already in possession of a workable piece of land. Both expropriation and organization implied, to many, an open turn towards socialism; and the traditional concomitant of such fears was the flight of capital. As private capital flowed out of the country, international borrowing became the only way to obtain the considerable resources required to set new development programmes in motion. But borrowing increased indebtedness, which led to monetary instability; and monetary instability hurt everyone.

The government of Luis Echeverría (1970–76) confronted this dilemma and, like the Cárdenas administration, was overwhelmed by it. Heavy, but inefficient, State investment in local rural development projects, a guarded opening of possibilities for independent peasant organization, and a limited number of expropriations of large properties held by influential members of the official party ceased abruptly with the political and economic crisis of August 1976, in which the peso was devalued by almost 100 per cent and rumours of a *coup d'etat* circulated freely. After that experience, any idea that agrarian policy in Mexico might soon again be associated with an open challenge from the State to the power of the post-revolutionary capitalist élite of the countryside was doubtful at best.

THE FUTURE OF LAND REFORM IN MEXICO

Land reform nevertheless continues to be the central issue around which the majority of the rural population of Mexico organizes its demands for assistance in meeting the basic requirements of livelihood. This is not an atavistic reflection of peasant nostalgia; it grows out of a realistic assessment of a situation in which employment possibilities outside agriculture are insufficient to provide a subsistence living, and access to land is therefore a crucial element in survival. In the early 1970s, most of the roughly 2.5 million landless labourers and a good proportion of the 1.5 million holders of infrasubsistence and subfamily plots in the country placed their hopes for future improvement in levels of living squarely on the possibility of securing a more adequate piece of land—in some cases, with the idea of growing a few commercial crops and, in others, with the hope of concentrating on subsistence production. Like their forefathers who fought in the revolution, they were, in general, engaged in a desperate effort to remain or become peasants, and thereby to assure some control over the conditions of livelihood at a time when the overall trend was clearly toward the proletarianization or semi-proletarianization of most people in the countryside.

After sixty years of land reform, and the distribution of slightly more than one-half of the cultivable land of the country to peasant beneficiaries, there is not sufficient remaining land to meet the needs of these new petitioners entirely. There are, however, enough clearly recognizable regions of reconstructed *latifundios* to make the continuing pressure for land reform an explosive challenge to the stability of the political system. If only token concessions are to be made to peasant attempts to force redistribution, as the failure of the very mild Echeverría programme would indicate, then the agrarian and agricultural problems of Mexico may well be attacked through a reorganization of the countryside from above, by a State cooperating closely with the large landholding, commercial, and banking élite of the nation, and holding out to masses of discontented rural people the promise not of land, but of employment.

The organizational vehicle for this new strategy (widely discussed by 1978) would be the 'associated' agricultural enterprise, established with the capital of private entrepreneurs or that State on land ceded by smallholders, who become shareholders and paid labourers. Management becomes entirely the responsibility of technicians. The enterprise is likened by its proponents to the collective and cooperative experiments of the Cárdenas years, in which the scattered resources of the peasantry were grouped in some cases into more productive units, which provided a higher standard of living to participants. But there is, in fact, a fundamental difference between the programme carried out during the 1930s and the present one: the balance of political forces in the earlier period permitted the exercise of considerable power by the peasantry, but that of the 1970s does not. The promotion of 'association' between landless labourers or *minifundistas* and private or public capital thus seems to imply the real risk that only those jobs would be created, and only those crops grown, which meet the financial needs of investors. Those needs, in a relatively free market economy, are not congruent with the basic livelihood requirements of most rural people. Therefore, while some 'associated' peasants might obtain a higher standard of living through participation in the new programme, most in all probability would not. A further extension of proletarianization in the countryside might well follow upon the end of innumerable informal arrangements through which landless labourers have gained access to small plots for self-provisioning, and whatever vestiges of local control over livelihood still remain might be eliminated.

Peasants in a number of regions have therefore opposed association with private or State capital in the founding of agricultural enterprises which convert former small landholders into hired hands and shareholders, and which make landless family members and neighbours superfluous to the productive process. If this is indeed to be the groundwork of future agrarian policy, there is good reason to suspect that it will encourage organized peasant protest, some of which may well be violent. Until a better way is found to meet the real material and social needs of low-income rural families, control over any plot of land—however small—will not easily be relinquished.

Environment, Society, and Rural Change in Latin America
Edited by D. A. Preston
© 1980 John Wiley & Sons Ltd.

CHAPTER 3

Agrarian reform in Peru

CLAUDE COLLIN DELAVAUD

1. INTRODUCTION

In 1964, the agrarian question was the dominant topic of political and social interest in Peru. It was, however, a long-standing factor in the social, intellectual, and political life of Peru; ever since the period of the Vice-Royalty, and throughout the nineteenth century, agrarian riots have broken out here and there in the sierra, as seen in the almost universal uprising of the peasantry in the south in 1780–82, led by Tupac Amaru II, forty years before the Independence.

The literary and political intellectual movement of the end of the century, and of the pre-war period, spearheaded by Mariategui, opened up the agrarian question and considered compensation for the long-standing acts of injustice perpetrated against the indians. The acceleration of the destruction of indigenous communities and the springing up of agro-industrial plantations from 1860 was also much discussed. But the period between the two wars saw, particularly on the coast, the trade unionism behind the APRA party face up to the double problem of the proletariat of the sugar plantations and its relationship with the small farmer sectors of *minifundio* which supplied the temporary workforce.

After the Second World War, the agrarian question was seen less in terms of justice than as an economic and social problem. Agricultural reform, supported not only by Marxist circles but also by the Christian Democrats, had four distinct roles. The humanistic goal of land belonging to the man who cultivates it, in reality, involves several methods of apportionment and has several aims. The most obvious of these is to give back land to all those people who have lost it and have worked as regular workers, sharecroppers, and day labourers. They are to become owners of their land, individually or collectively. The division of the mismanaged plantation into small farms will increase productivity and favour the growing of cash crops, and reduce the produce available for export. The third objective concerns the general economy. The small, independent peasantry will constitute a new market, internal and expanding. Lastly, the landowning oligarchy will be deprived of one of its traditional controls and will be obliged to turn its attentions towards the secondary and tertiary sectors.

Agrarian reform is seen first of all as a modernization of agriculture within the framework of a more independent economy and as a means of the national integration of the rural masses, especially the mestizos and indians of the sierra. Gradually, the revolutionary and reformist political parties adopted its principle between 1955 and 1962, as well as the University and an increasing number of members of the Church and the Army. After the Presidential elections of 1963, all the major political schools of thought and sections of the nation expressed the wish for agrarian reform, but obviously not the same reform for everybody. For the Christian Democrats, backed by members of the University and the left wing of the Church, it must be radical; that is to say, it must completely reshape man's relationship with the land. In place of the oligarchal agrarian structure (which, in 1961, among other things, left 11.3 million hectares in 1091 very large estates of more than 2500 hectares, i.e. 61 per cent of the land for 0.1 per cent of owners—see Table 3.1), the intention was to substitute a system of cooperatives open to everyone who works the land. The reformists of the Acción Popular with the future President Belaunde and the old populist party APRA of Raúl Haya de la Torre envisaged an agrarian reform which would abolish indirect exploitation and advance the indigenous peasant communities, starting with the existing *minifundio* and the confiscated large developments.

For the conservative sectors, represented by General Odría, agrarian reform would only concern the land of those *latifundia* devoted to extensive stock raising and those badly managed lands of partial agricultural occupation. The difference between these outlooks is enormous, and the misunderstanding will continue for a long time and still exists today.

Table 3.1 General agrarian structure in Peru in 1961

Farm size (ha)	Number of holdings		Total area of farmland	
Less than 1	290 900	34.2%	127 869	0.6%
1–5	417 357	48.9%	926 851	4.9%
5–10	76 829	9.1%	481 631	2.6%
10–20	30 370	3.6%	397 754	2.1%
20–50	17 414	2.1%	506 745	2.7%
50–100	7 214	0.8%	474 313	2.5%
100–200	4 606	0.5%	589 567	3.2%
200–500	3 475	0.4%	1 035 076	5.5%
500–1 000	1 585	0.2%	1 065 157	5.7%
1 000–2 500	1 116	0.1%	1 658 639	9.3%
More than 2 500	1 091	0.1%	11 341 901	60.9%
Total	851 957	100%	18 604 500	100%

Source: Agricultural Census of Peru 1961.

2. THE AGRARIAN QUESTION IN PERU ON THE EVE OF THE 1964 AGRARIAN REFORM

2.1 A very unequal distribution

Originating from the right of conquest of the Spaniards since the sixteenth century, then the ceaseless encroachments of the colonial *haciendas* in the two following centuries, and, finally, the formation in the nineteenth century of *latifundia* devoted to stock raising in the sierra and of the large sugar and cotton plantations of the coast, the Peruvian agrarian structure is typical of an oligarchal occupation of the land, where two-thousand farmers of more than 1000 hectares appropriate more than 70 per cent of Peruvian soil. But the remaining 30 per cent is equally unfairly distributed, since 83 per cent of the farmers develop scarcely 5 per cent of the land, thus forming a sector of peasants with very small holdings less than 5 hectares in size. At the very bottom of this scale, 34 per cent (i.e. 290 900) of the farmers cultivate even less than 1 hectare. Between these two extremes can be found both a category of farms averaging 50 to 1000 hectares, which occupy, with 1.9 per cent of the total manpower, 16.9 per cent of the surface area, and a category of 'family' developments of between 5 and 50 hectares, where 14.8 per cent of peasants work no more than 7.4 per cent of the land (see Table 3.1).

Thus, between the large landholders and the mass of smallholders, the basis of life of the peasant family or middle-sized farmer appears very weak. However, the Peruvian agrarian question is made more complex because of the different geographical environments on which Peru is based, as well as the interplay of the different social sectors which make up the peasant class.

2.2 The three different Peruvian environments and their agrarian consequences

Peru's structure is simple, but strongly contrasting. From west to east one can distinguish three major ecological zones, the coast, the sierra, and the *selva*, which are now considered.

2.2.1 *The coast*

The large piedmont zone of the Pacific side of the Andes comprises the Peruvian coastal region, whose extreme dryness has made it into a desert, and which stretches from the extremely dry Chilean coast to the border of Ecuador, where the semi-arid conditions typical of the Sahel can be seen in their extreme. But the rivers, fed by the rains and the glaciers of the cordillera of the Andes, have formed twenty-two valleys where sufficient water flows permanently or at least during the southern summer to allow irrigation. For more than 3000 years, the natives of the coast have changed these valleys into oases, constructing an entire system of irrigation, taken over by the Spanish conquerors and then modernized since the end of the last century.

Here, the fundamental problem is not land, but water. The indians cultivated about 800 000 hectares but, since the nineteenth century, the plantations of sugar-cane, rice, and cotton which consume more water than the indigenous food crops, have reduced the irrigable areas to approximately 500 000 hectares. The motor-driven pumps of the twentieth century have enabled an increase in output in the sugar plantations of the northern central region, and a gain of some 100 000 hectares of cotton in the north but, with 600 000 hectares under farms, the coast reached its peak in the 1960s. Worse still, the modernized, highly capitalized, and industrialized plantations which replaced the colonial *latifundio* of the coast have, since the second half of the last century, made numerous inroads into the neighbouring indigenous communities. There has either been misappropriation arising from the vagueness of the title-deeds, or else the reserved lands have been sold by careless or corrupt *caciques*. The rights to water have been transferred, along with the lands.

The farmland of the plantations is, as a general rule, situated upstream, i.e. at the water catchment area, the average estates of colonial origin surround the towns at the centre of the oasis, and the peasant communities which have escaped the various stages of encroachment only continue to exist downstream. They benefit only from what water is left once the upper region has been served or from ancestral rights which have lost the major part of their practical significance. Finally, the massive increase of the water supply to the large plantations have led to the leaching of land upstream and overenrichment downstream. The incidence of salinization from Libertad to Piura has, since 1930, begun to ruin much land belonging to the indigenous community downstream, whose too-salty soil had become unusable. If one bears in mind that the demographic explosion started shortly before the Second World War, one will appreciate that the peasants' situation on the coast became tragic from the end of the 1950s. In 1961, out of the 600 000 cultivated hectares of the coast, 120 000 hectares were cultivated by the nine large sugar producers which have absorbed 62 *haciendas* within a century. Also, 190 000 hectares are cultivated by 195 cotton plantations, and 110 000 hectares by 320 farms producing food, mainly rice and maize, i.e. produce easily cultivable and negotiable. The rest, some 180 000 hectares, is cultivated by smallholders who devote themselves to cultivating a variety of foodstuffs; more than three-quarters of them are self-sufficient. Their number is increasing and the area under cultivation is decreasing, because of the deterioration of the irrigation system and the increasing salinity downstream. The *minifundio* system gets worse from south to north, and reaches its height at the lower part of the Piura valley with 60 m^2 for each tenant farmer's family.

To combat this situation, major irrigation schemes for these parched coastal areas have been initiated. That of San Lorenzo made it possible to develop 48 000 hectares of new land. Not without disappointments during the dry years, the

Tinajones scheme was designed to develop 90 000 hectares of the oasis of the Chancay and Leche valleys which was to be completed in 1976. The River Santa scheme will involve 120 000 hectares in Libertad, and the Olmos scheme will make it possible to irrigate a new area of almost 100 000 hectares. In the south, the development of the Majes canyon anticipates the recovery of 60 000 more hectares. At Tumbes, in the extreme north, there are another 30 000 hectares which may be irrigated. On the Jequetepeque, another 40 000 hectares can be irrigated. None of these schemes, however, has been completed in 1978.

Only the diversion of water from the Chira towards the Piura, by means of the Poechos dam, has made it possible, since 1975, to irrigate finally the 60 000 hectares of the Lower Piura and to irrigate the 40 000 hectares of the Chira valley.

Back in 1961, apart from the San Lorenzo dam, nothing had yet been built and there was a shortage of water everywhere on the coast, where water rights in a dry year, i.e. four out of five years, are purely theoretical.

2.2.2 *The sierra*

The highlands comprise a discontinuous area between the coast and the Amazon basin. The land, provided that it is not too sloping, can be exploited by *secano*, or dry farming, dependent only on the rains. The output can be improved by irrigation, which is possible in the *hoyas*, or interior basins, but it is not the general practice, even for the *haciendas*.

The sierra is the traditional domain of stock-raising *haciendas*. Sheep rearing predominates throughout the south and the centre, cattle rearing in the north towards Cajamarca. The land is shared between the traditional *latifundio* of colonial origin (extensive and often badly managed by absentee landlords, except in some basins of the north and the centre) and the remaining indigenous communities. The *haciendas* themselves are divided into two sections, one of which is administered directly and the other operated as an indirect exploitation by sharecroppers. The chief problem here is no longer water, but the huge area. By the side of the insufficiently cultivated, enormous *latifundia*, which can exceed 100 000 hectares and even be as much as 350 000 hectares, the 280 000 tenant farmers *(pongos, yanaconas, aparceros, colonos*, and *partidarios*, grouped together under the common denominator of *feudatarios*) possess scarcely enough land to support a family. They have to carry out additional work on the employer's own property in payment for the right to occupy land, but are devoid of any security of tenure. It is much the same for more than 400 000 peasants living in communities on the poorest lands. In fact, the *haciendas* have occupied either the cultivable basins or the favourable slopes, the terraces or the large, flat stretches of the *puna*, where they rear livestock. The communities are isolated on the steep slopes, in geographical sectors far removed from the axes of communication, or else on slopes with very poor soil.

Here, too, there is a shortage of space for the sharecropper, for the 680 000 tenant farmers of the sierra are *minifundistas*, with regard to the mediocrity of their land and as a consequence of the techinques allowed by their situation. They have also to supplement their income by going to work temporarily on the *haciendas*, and in addition they are forced to accept the emigration of some of the surplus population.

2.2.3 The "selva"

The Amazonian basin of Peru still remains very sparsely populated. There is no shortage of space or water. That remains academic. In fact, the best lands, those of the Ceja de la Montaña, or upper piedmont, which are well drained, are monopolized by coca and coffee *haciendas*. There are certainly some areas still not cultivated, but they are owned by the *haciendas* or by village communities. In the extreme south, the Aymaras have steadily penetrated and colonized at the same time. In the south, towards Cuzco, the valley of La Convención has been occupied by fairly important landowners who have attracted both sharecroppers and labourers. From 1962 onwards, these were the cause of serious agrarian disturbances. In the northern central region and in the north, old colonies, San Martín, or new, Jean-Bagua, suffer from a shortage of infrastructure and equipment and have made no progress, or even slip backwards, as in San Martín. Scarcely 150 000 people live on these lands, which have more often than not been abandoned to diseases and technical neglect, in the face of the harshness of the equatorial environment.

Amazonia did not seem, in 1961, to confirm the hopes of those people who would like to deposit there hundreds of thousands of families who have no land, or at least are not the owners of their land.

3. THE DIFFERENT SOCIAL SECTORS OF THE PEASANTRY

Several splits can be distinguished in rural Peruvian society. The average and large traditional estates of the coast have gone the same way as the large-scale capitalist development. Even if they have not been modernized through a lack of funds or of initiative, they have been impelled by the large estates to produce sugar, cotton, rice, and maize by the integration of their system of production and commercialization, often leaving part of the mechanized work in the hands of the large estates. Study of sample families shows that one passes often from renting the land, worked by others, to pioneer development, as at Ica or in Piura, where *hacendados* own between 60 and 300 or 400 hectares. In the Sierra, the medium-sized estate is rare still. As for the traditional *hacienda*, it remains characterized by low productivity and indirect exploitation.

In fact, the peasant population, far from constituting a homogeneous class with a relatively similar way of life and income, is made up of very different social groups, in the sierra, the coast and Amazonia. In these areas, one finds in fact, to a greater or lesser degree, permanent agricultural workers, real proletarians, employed on a monthly basis, enjoying certain guarantees of employment, social security, and the right to belong to a trade union, all this in establishments employing at least twenty people.

Alongside this group, which is especially important in the large coastal plantations and certain huge stock-raising estates of the sierra, one finds the category of indirect tenant farmers or *feudatarios*. These are sharecroppers, settled on the outskirts of an estate, holding in usufruct a farm which is usually sufficient to support a family. These tenant farmers pay a rent linked to the harvest, of anything between one-third and one-fifth of the harvest, and owe one or two days of work.

These two groups, permanent workers and sharecroppers, mostly coexist in the traditional *haciendas*, but also in most of the capitalist plantations, where they constitute a reserve workforce and a not-inconsiderable source of revenue in periods of recession in the export economy. At the two extremes, besides these groups based on the *hacienda*, are the landowners, members of freeholding communities, and day labourers. The former, with 85 000 families, constitute scarcely 9 per cent of the peasantry, the small and middle class of direct cultivators, ranging from the family tenure to the medium-sized traditional *hacienda*, often called a *fundo*. But the great mass of small landowners, 551 900 families, are registered as member of freeholding communities (see Table 3.2). It is not always a question of native communities, duly recorded as such, but also of village communities, sprung from villages of *hacienda* sharecroppers, which have become independent communities since the middle of the last century. Whatever the legal nature of these communities, it never implies collective or cooperative agriculture. Cultivation is by the individual family, and only the unalienable nature of the land confers some protection on the native community and its members. Some community practices, especially of mutual assistance and sometimes of redistribution of land for less-privileged families, exist in the central and southern sierra, but cultivation remains essentially individual, and fundamental inequalities are commonplace: some have ten times the area of others.

Table 3.2 The status of peasants in 1961

	Agricultural population: number of families		Surface area	
Landowners	85 000	8.2%	16 364 244	87.7%
Comuneros	551 900	52.9%	2 240 256	12.3%
Landless	404 813	39.2%	—	—

Source: Agricultural Census of Peru 1961.

Generally speaking, in the coastal communities of irrigation agriculture, three-quarters of the families are smallholders and a community spirit no longer exists, except for the collective organization of defence in the face of the encroachments of the *haciendas* or neighbouring plantations, or malpractice by those who distribute the water.

In the sierra and certain warm valleys of the piedmont zone of Amazonia, communities are not, strictly speaking, smallholders, but marginalization either geographical and technical, or cultural, syndical and political, brings about the same economic consequences as does ownership of little land.

All the small landowners and smallholding and marginal sharecroppers have to find supplementary work in the *haciendas* and capitalist plantations. These day labourers or *braceros*, do not benefit from any guarantee of employment, do not get social security, nor any support from the unions. Their actions sometimes conflict with those of the permanent workers or of the sharecroppers of the estate, where they come to work only at harvest time or to prepare the soil. This category exists especially on the large coastal plantations or in certain estates of the montaña of the Amazon basin.

Finally, with a demographic explosion coming directly after the second phase of the large estate's encroachments upon the lands of the peasant communities, these have been in straitened circumstances since the period between the two wars. As the surplus population cannot be absorbed, 20 per cent of the peasants of the coast have no land nor regular employment, thus constituting a huge reserve of manpower for the coastal plantations. As for those of the sierra, the very poor quota of daily jobs provided by the stock-raising *haciendas* means that they have to emigrate towards the coast, usually towards the large towns.

4. SOME AGRARIAN REFORMS AND THEIR CONSEQUENCES

4.1 The reform of 1964

In 1964, President Belaunde succeeded in pushing through, with a hostile majority, a scheme described as reformist. It was, in fact, a compromise between the aim of a radical redistribution and the opposition of the large landowners. It stipulated that the landowners on the coast could keep 150 irrigated hectares and on the sierra 300 hectares devoted to dry farming, and between 600 and 1000 hectares pasture land. The remainder of the land would be given to the regular agricultural workers, to the tenant farmers, and, finally, to those peasants who had no land.

Article 15 especially conferred land cultivated indirectly to the serfs, farmers, and sharecroppers who became landowners. Only this provision was of real importance, as it created a new class of landowner. It could improve family productivity without, however, destroying the structures of production, since the sharecroppers were not obliged to form cooperatives immediately.

Two provisions were to restrict even more the extent of the application of the law. The first excluded the agro-industrial enterprises from agrarian reform. That amounted to the exclusion of all the large sugar plantations concentrated around nine large sugar refineries on the coast to the north of Lima. A large cotton estate attached to an oil-works and soap-works at Piura was considered to be in the same category. Failure to prevent the *haciendas* being shared between stockholders or heirs in the same family made it possible, moreover, to go ahead with legally breaking up the large estates and capitalist plantations into farms of 150 hectares whenever possible, which were then exempt from the reform, but were now no longer effective production units.

In 1968, 324 large, but geographically marginal, estates, almost all situated in the sierra, were dispossessed, and 12 000 families received a share of land, as sharecroppers of *haciendas*, under Article 15. Agrarian reform was applied case by case, following an obviously lengthy procedure, since no land survey was then in existence, and many of the tenant farmers did not possess any title-deeds. Finally, the departmental records of the application of this reform suggest that, judging by the performance of the first four years, it would take a good 15 years to conclude the case.

4.2 The agrarian reform of 1969

The revolutionary Military Régime installed in October 1968 changed both the speed and the scope of the agrarian reform. The new Régime intended to overhaul the structure of Peruvian society and economy in every sphere, giving priority to that prevalent in rural areas throughout Peru. The new agrarian reform of June 1969 envisaged the integration of the rural masses, both indians and mestizos, into the Peruvian population. The double objective of national integration and the independence involved the elimination of the landed oligarchy and the formation of a homogeneous peasant class. The sierra and the Amazon lowlands were to be integrated with the coast, the country's economic and sociological centre of gravity. The indian rural masses of the sierra and the mestizos of the coast were to be integrated; the new farmers were to be introduced into the Peruvian market economy. At the same time as the landed ruling class is destroyed, a new agricultural power, both social and political, must be created. The peasant was to be demarginalized geographically, culturally, economically, and socially.

The new reform no longer makes any exceptions. The huge capitalist agro-industrial complexes are affected. From March 1970 onwards, the *haciendas* can no longer be subdivided between relatives or shareholders, nor voluntarily be portioned out for the benefit of sharecroppers or workers. Finally, the size limit of 150 hectares on the coast would no longer be observed after 1972, if the administration has been unsound, and if the social laws or the terms of work and accommodation have not been adhered to. All the capitalist plantations and the *haciendas* would be practically affected from 1972 on the coast and expropriation would be under way after 1974.

The process was longer and less smooth in the sierra. But this time agrarian reform was intended to advance not so much a class of small landowners as a self-directing society. The intention was to transform the modernized capitalist plantations, like the large estates, into production cooperatives. The small traditional *haciendas* and the more recently created *fundos* of average size will be grouped together to form a cooperative.

4.2.1 Consequences on the coast

The nine large agro-industrial sugar complexes were first of all controlled by the State and then handed over to the workers in 1971 in the form of self-administered cooperatives. In theory, the State, by means of a sugar federation, will strictly control the general strategy of production and investments. It is, however, the technical and financial cadres who manage the complexes, which are unwieldy and difficult to administer especially because all the essentials for cultivation apart from the labour force (such as very mechanized equipment, manure, pesticides, and energy) belong to the State, just as the sale price of the sugar will also be dictated by the State. For the cooperatives producing cotton, rice, and other foodstuffs, the dualism between appointed cadres and elected councils is often very marked, but the State leaves the workers a lot more initiative. The strong influence of trade unions of the APRA movement or those of the Marxist agrarian league CCP after the council elections led to masked or open conflict with the bureaucracy of SINAMOS (Sistema Nacional de Mobilización Social) entrusted with the leadership of the cooperative movement.

Almost all the *haciendas* on the coast were appropriated, then expropriated, and finally transformed into cooperatives. Even those of less then 150 hectares were unable to survive this period without being appropriated, or invaded and then appropriated. This was particularly common at Piura and Lambayeque, but occurred in almost all coastal areas. The sharecroppers and the workers became full members of the cooperative. If there was more than 4 hectares of land (together with its water right) for each legal member of the cooperative, landless peasants were admitted. In 1974, this quota was even reduced to 2.5 hectares. In fact, agrarian reform on the coast gave the land to those people who already had regular work or to the indirect tenant farmers; of the mass of the 200 000 landless peasants, no more than 10 per cent were affected. Many, moreover, were related to cooperative members, as sons, brothers, or uncles, etc.

It was impossible to admit to the cooperatives the mass of day labourers, whether they were landless peasants or smallholders. For these latter, too, agrarian reform did not realize their expectations. The deprived village communities were not, generally, able to occupy the coastal *haciendas*, which have now become cooperatives, as they were prevented by the workers or sharecroppers who had become cooperative members upheld by the law.

Agrarian reform did not relieve congestion in the smallholdings of Piura and Lambayeque in the north, nor those of Arequipa in the south. Even within the affected *haciendas*, there was antagonism between the workers who became cooperative members of the estate controlled directly by their former employer, and the sharecroppers who wanted to keep and cultivate their farms individually. It was necessary to compromise with them, and eventually most of them joined the cooperatives and retained all or part of their farm.

In 1975, it seemed obvious that only those formerly provided with employment or land by an estate owner would benefit from agrarian reform, not without clashes with bureaucrats and technicians, nor without about 28 per cent of bankruptcies in 1976 in Piura and Lambayeque alone. On the other hand, 20 000 (i.e. 10 per cent) of the landless peasants joined cooperatives. The majority of the smallholders (80 000 families) and of landless peasants (180 000 families) remained outside the system.

4.2.2 Consequences in the sierra

All the *haciendas* of the sierra displayed some characteristics of large estates. Stock farming was predominant, the sierra only produced 25 per cent of foodstuffs and yet there was very little cultivation of produce for export. Certainly, some capitalist stock-raising societies had been created in certain basins at Cajamarca, in Junín, and even in the southern central region, but nine-tenths of all estate lands were exploited extensively.

The proportion and the total number of permanent workers was much less than on the coast. On the other hand, the indirect tenant farmers (*pongos, yanaconas, aparceros, colonos*) were proportionately more important. Alongside the *haciendas*, village communities are more numerous and more powerful than on the coast. Indigenous communities more often than not are united historically, culturally, and socially. With a long tradition of strife, the support of the State, sometimes dating back to the colonial era, but especially after 1920, was theoretically guaranteed to them. They influenced the development of agrarian reform much more than on the coast. It is they who, by their action in 1969, at Huanto and Ayacucho, precipitated the new agrarian reform.

Paradoxically, the large estates (although badly run by absentee owners paying no attention to labour legislation, unproductive, condemned by practically every political group, and abandoned by the first reform) resisted agrarian reform more successfully than the capitalist *haciendas* of the coast. There are technical reasons. The *hacienda* of the sierra is often smaller, very badly surveyed, and geographically very badly defined. There are also legal reasons. It is often engaged in law-suits with one or several communities, and involved in litigation with its neighbours, and also its sharecroppers.

There are also sociological reasons. The environment of the *hacendados* of the

Table 3.3 Progress of the Peruvian Agrarian Reform at 31 July 1976

| | \multicolumn{13}{c|}{Agrarian zone} |
	I	II	III	IV	V	VI	VIII	IX	X	XI	XII	XIII
Total allocations												
Area (1000 ha)	640	461	1107	130	132	97	29	5	1394	854	1253	341
Families (1000)	28	31	51	18	12	3	1.3	0.7	39	46	21	28
Farms	111	155	135	96	78	14	8	4	181	283	46	160
Individual allocation												
Area (1000 ha)	4.7	10.2	7.6	9.7	5.3	6.9	4.3	—	4.6	78.5	5.8	—
Families (1000)	2.6	1.2	1	2	1	1.4	0.8	—	0.3	8.9	0.4	—
Cooperatives												
No. of cooperatives	56	83	38	82	67	6	5	4	40	59	10	28
Area (1000 ha)	326	247	118	60	77	29	1.6	5.2	232	412	469	120
Families (1000)	10	23	15	14	9	0.6	0.2	0.7	3.6	13.6	6.3	4
Groups of former tenants												
No. of groups	41	67	53	2	2	5	—	—	58	162	2	103
Area (1000 ha)	237	110	99	0.3	87	6.4	—	—	205	242	25	111
Families (1000)	3.2	5.2	6.1	0.07	0.4	0.1	—	—	3	7	0.05	6.4
New communities												
No. of communities	14	3	22	11	9	1	2	—	77	55	14	27
Area (1000 ha)	36	31	36	44	63	0.2	4.3	—	324	73	8	33
Families (1000)	12	0.6	4.7	1.3	1.7	0.4	0.08	—	23	13	2.7	7.4
Sociedades Agrícolas de Interés Social												
No. of SAIS	640	2	22	1	—	2	—	1	6	20	20	2
Area (1000 ha)	—	629	846	14	—	53	—	19	628	48	845	77
Families (1000)	—	0.7	24	14	—	0.4	—	0.2	9	2.8	12	10

Source: Agrarian Reform Institute.

sierra is often more closely linked with the local population, by birth and by cultural origin. Whereas a great many *haciendas* on the coast belonged to creoles, often allied to foreigners, the traditional *haciendas* of the sierra, especially the small and medium-sized ones, date back a long time, having been set up by well established creoles and often by mestizos. These did their utmost to exert pressure on urban society, but also on the officials responsible for agrarian reform.

This reform was first applied to the obvious socially important estates, better surveyed and impossible to defend. But, for the others, technical delays and legal confusion will delay, if not the appropriation, at least the expropriation, and especially the third phase of adjudication. Technical and sociological delays brought about veritable confrontations in the central and southern sierra. As for the northern sierra, where the stockbreeders of Cajamarca modernized their *fundos*, which are considerably less in surface area than in the south, it was left alone until 1973.

By 1976 (see Table 3.3), all the large estates were finally affected and expropriated, but at the cost of a double struggle in the centre and the south. The former set the SINAMOS with its ally the CNA (Confederación Nacional Agraria) against the CCP (Confederación Campesina del Perú) formed by trade unions, largely Trotskyist and Maoist in inspiration. The second struggle highlights the internal contradictions of agrarian reform, by bringing the new cooperatives formed by one or more *haciendas* and run by former workers and sharecroppers into direct confrontation with the *comuneros* from the neighbouring indigenous communities. These were either victims of long-standing injustices and often involved in litigation with the neighbouring *haciendas*, or, at the very least, they owned insufficient land for their members who worked as day labourers without this enabling them to become cooperative members. In Ancash and Junín in the centre, in Ayacucho and Huancavelica in the southern central region, in Apurímac, Andahuaylas, Abancay, Cuzco, and Puno in the south, there was permanent conflict between 1972 and 1975. From that date, it was resolved very slowly and half-heartedly.

In 1976, in the sierra, out of 960 000 rural families, 226 000 (i.e. 23 per cent) received some land either as individual parcels or as cooperatives. This amounts to saying that only the hard core of the *haciendas* were handed over by agrarian reform to the workers or sharecroppers. The remainder, peasants in communities or landless day labourers, received nothing. The expropriations affected eight million hectares and a little less than seven million were transferred.

4.3 The attempted integration of the cooperative and community sectors

The summary of the restructuring of agriculture in 1976 appears clear. Agrarian reform turned 11 664 private farms, representing 8 million hectares, into 1271 cooperatives, covering a little less than 6.8 million hectares. The remainder was to be turned into about 150 other cooperatives; in effect, once it was totally completed, agrarian reform would have grouped the small, medium-sized, and large *haciendas* into self-administered cooperatives, generally eight times larger than the average *hacienda*.

The average size of farms prior to 1968 of 686 hectares grew, as the agrarian reform process drew to a close, to 5350 hectares. This national average farm size applied particularly to the sierra and to the eastern lowlands, where arable land comprises only 8 or 9 per cent of the total area. Farms in these areas seldom supply more than 25 per cent of Peruvian food production. On the coast, the average size is much less, about 300 hectares for the cooperatives formed from small or medium-sized properties, but, in the irrigated farmlands, the output is from 400 to 800 per cent more than the dry farming of the sierra. The coast supplies 75 per cent the nation's food and 90 per cent of the nation's agricultural exports. This produce for export can be grown just as well by the small cooperatives as the large ones, but in lesser quantities. The small cooperatives formed by amalgamating the small *haciendas* or coastal *fundos* specialize primarily in growing rice in the north, maize in the centre, and intensive dairy farming in Arequipa. On the other hand, the large capitalist plantations of the centre and on the north coast have become enormous cooperatives, corresponding in size and the way in which they are administered to the former plantation, in order not to disrupt the system of production.

Thus, very great differences still exist on the coast between the agricultural enterprises. The nine sugar cooperatives have each between 7000 and 25 000 hectares of irrigated land, from Lima to Lambayeque, and the twenty-eight large rice or cotton cooperatives each have more than 1000 hectares. The 203 mixed cooperatives, growing rice, cotton or vegetables, have between 160 and 600 hectares each. The preservation of existing structures for the cooperatives formed from the very large capitalist agro-industrial developments has made possible, despite social friction, and with only one exception near Cayalti in Lambayeque, the maintenance of production and even of productivity. It has not been quite the same for the large rice and cotton cooperatives, whose future has varied according to the size and previous organization of production. Productivity has been kept at the same level with difficulty and production has decreased, by at least 10 per cent, in more than half of them.

It is in the small cooperatives where a dictatorial restructuring, often poorly backed up because of the lack of competent trained staff and failure to secure credit, that social conflicts have broken out, and bankruptcies followed by a request for State intervention have been frequent. More than 40 per cent of them have come up against serious difficulties, forcing the State to suspend its demands for capital and tax annuities. The greatest problem on the coast and in the sierra is, above all, the widening of the gulf between the cooperatives formed from the traditional large estate or capitalist developments and the communities of small independent peasants. By June 1976, agrarian reform had affected 279 000 families and, by the end of 1978, it will have affected 340 000 out of about 1 200 000 peasant families. This means that 72 per cent of the peasants have not been affected, scarcely half of whom are small landowners and the other half landless peasants.

A consequence of this has been the attempt to integrate at least the peasants of the village communities. Two systems were introduced: the PIAR and the SAIS. In the face of the internal contradictions of agrarian reform, which set production cooperatives, badly structured groups of former sharecroppers, independent smallholders, indigenous communities, and temporary landless workers against one another, PIAR (Proyectos Integrales de Asentamientos Rurales) were established in April 1973. They bring together over a very wide area, in the region of several thousand irrigated hectares on the coast, different cooperatives producing export goods, groups of ex-sharecroppers growing crops, and temporary landless workers who had not been able to join the cooperatives. Each PIAR functions as a single enterprise, with its own land, its own equipment, and its labour force, with no distinction of origin. In theory, it should even integrate the sectors of small landowners, but the number of these and their refusal to integrate their lands into the PIAR has up to now delayed their introduction. On the other hand, some smallholders have joined PIAR as former temporary workers. In the whole of Peru, especially on the coast but also in the sierra, in the regions where farming as well as stockraising is practised, 52 PIAR were theoretically established starting in 1973, grouping together 467 cooperatives or groups of farmers. The restructuring will be very slow. It is necessary not only to break up the original production units but also to integrate new workers. That can only be done by a fundamental modification of the production system, additional technical expertise, and huge credits, that is to say a general State intervention.

The PIARs, theoretically self-administered, come under tight State control. Only about twenty are in operation on the entire coast, largely in the north. The sectors of small landowners and irrigators still exist in Piura and Lambayeque in the north, and Ica and Arequipa in the south.

The SAIS (Sociedades Agrícolas de Interes Social) were largely conceived to integrate the very large stock-raising estates, with their regular workers and their sharecroppers, and also the *communeros* of the neighbouring indigenous communities, at whose expense the large estates were established. They can, like the Tupac Amaru SAIS set up on the former stock-raising territory of the Cerro de Pasco mine in Junín, occupy more than 500 000 hectares. Here, too, the State has to exercise fairly considerable control of mixed and self-administered societies, which it must supply with financial backing and expertise.

CONCLUSIONS

On the whole, in 1978, Peruvian agrarian reform has practically eliminated the entire former system of large landholding and capitalist production. The land belongs to the person who was cultivating it before the reform. It has also avoided aggravating the problem of micro-holdings by grouping together into production cooperatives the main body of beneficiaries, workers, and tenant farmers. But because of insufficient land in the sierra and inadequate irrigation water on the coast, it has been impossible for it to integrate more than 10 to 20 per cent of the landless workers. In the same way, the majority of communities of small irrigators of the coast and indigenous communities (which have now become peasant communities) of the sierra have not been able to be transformed into cooperatives.

After eight years of agrarian reform, the slow integration of the cooperative and small farmer sectors, started in 1973 within the PIAR and the SAIS, is very far from being completed. It will need time, strong regional outfits, credits, education, and faith.

Environment, Society, and Rural Change in Latin America
Edited by D. A. Preston
© 1980 John Wiley & Sons Ltd.

CHAPTER 4

Agrarian reform and rural change in Ecuador

MICHAEL R. REDCLIFT and DAVID A. PRESTON

AGRARIAN STRUCTURE AND DEVELOPMENT

The structure and performance of the agrarian sector in Ecuador can only be understood in relation to the country's economic history, which has given rise to the balance of social forces within both the coastal and sierra regions. Unlike some other Latin American countries, Ecuador is without an industrial economic base, and thus lacks well defined industrial classes with the capacity to enforce a policy of cheap foodstuffs on the dominant landowning interests. The political pressure for agrarian reform within the country has largely been mounted by the peasantry, in the teeth of persistent opposition from both highland and coastal landlords. However, in this chapter, it is suggested that spontaneous processes of market penetration within rural society provide a better guide to the form that rural development is taking than the provisions of agrarian reform legislation. In both major regions, private land sales have been more important than agrarian reform policy as a mechanism of land redistribution. In both regions, the general direction of agrarian development has been towards the capitalization of the estate structure, and the official encouragement of medium-sized entrepreneurs: both kinds of enterprise ensuring that wage employment plays a much larger rôle than in the past.

Outwardly, Ecuador exhibits the classic features of a 'dual economy'. The sierra is a region of extensive estates; the coast a plantation economy. In the highlands, the land tenure system has been the principal means of social and political control available to landlords. As in many other areas of the Andes, under pre-reform conditions, the landowning class of Spanish descent has effectively monopolized the best agricultural land. The independent smallholders, many of whom are of indian extraction, have been forced to eke out a living on inadequate holdings. Other peasant families have derived their livelihood from small parcels of land (*huasipungos*) within the large estates. This limited access to land was made in return for the *huasipunguero*'s labour services, usually so many days per month, to the estate owner and his family.

As population has increased in the sierra, the balance of communal resources within indian communities has been upset. Market pressures, particularly the need to earn cash through wage employment, and the sale of a rising proportion of their crops, have served to intensify the strains imposed by declining demographic/natural resource ratios. Simultaneously, some *serrano* landlords have been able to take advantage of market conditions to modernize their estates, adopt more extensive land uses, and employ limited numbers of wage labourers in place of sharecroppers or service tenants, on their estates.

Coastal society has also been marked by the domination of the large estate, but in this region it has taken a rather different form. Large estates were not important on the coast until the late nineteenth and twentieth centuries, when they were developed in response to world demand for tropical export crops. Consequently, the production of crops has followed successive cycles of 'boom' and 'bust', first cocoa, and then bananas providing the country's main source of foreign exchange earnings. Rural society on the coast came to reflect the patterns of land tenure and labour organization associated with export crop production. The trade stimulated by external demand also led to the development of Guayaquil, the principal port, as an *entrepot*.

Coastal export interests reached their zenith in the first two decades of this century, when the Banco Commercial *y* Agrícola *de* Guayaquil exerted almost complete control over Government economic and budgetary policy. After 1925, however, *serrano* interests were firmly re-established in the wake of a slump in cocoa fortunes. During the last half-century, an accommodation has been reached between coastal and *serrano* interests which, while it has blocked agrarian reform measures, has not prevented the development of capitalist farming enterprises in both regions.

AGRARIAN REFORM 1964–70

In 1960, Velasco Ibarra, a populist leader who held office intermittently for several decades, was elected to the Presidency of the Republic on the promise of agrarian reform. During the early 1960s, popular pressure for land disribution increased, and coincided with the interest shown in agrarian reform policy by the Alliance for Progress. By 1964, an Agrarian Reform Institute (IERAC) had been established in Ecuador, and the first Agrarian Reform Law had been passed.

It is interesting to compare the provisions of the 1964 Agrarian Reform Law with the analysis of Equador's agrarian structure provided in the report of the Inter-American Committee for Agricultural Development (CIDA, 1965). The CIDA report documented the relationship between land distribution and low productivity in the agricultural sector, and drew attention to the abuses to which this led. It was recognized that some highland estates were undergoing structural changes, as capitalist production and marketing relations became increasingly important. In the sections on the coast, the report emphasized the important part being played by families of colonizers in the banana-producing region around

Santo Domingo de los Colorados. Both processes were identified as beneficial 'modernizing' influences.

In its comments on the 1964 Law, the CIDA report suggested that the Law was 'more developmentalist than redistributive'. Reform was urged in the interest of 'modernization' rather than equity (CIDA, 1965, p. 497). Certainly the 1964 Law envisaged the maintenance of large estates, provided that they were actively worked by their owners. The Law had clear liberal credentials: it was inspired by respect for private landed property and the much-publicized need to increase agricultural production. Most of the legal provisions contained in the Law which required land to be worked efficiently were ignored in the years that followed. Only the abolition of the *huasipungo* and *arrimado* systems were implemented. However, as we shall see, the importance of labour service within the estate systems had already declined, and the 1964 Law, if it had any effect in 'freeing' land and labour markets, was only further strengthening existing processes at work in the rural economy of the sierra.

During the nine years before the next Agrarian Reform Law (1973) was passed, land redistribution increased in both coastal and sierra regions, although the Land Reform Institute was more active in encouraging colonization in the eastern jungle than organizing a 'reformed sector' in the sierra. On the coast, particularly, agitation for land by tenants created pressure for land sales, which were originally effected by utilizing private sources of credit. However, large plantations continued to be owned and managed by foreign companies, and the more economically 'efficient' agricultural enterprises were in no way affected by agrarian reform legislation. When the pace of agrarian change quickened after 1972, and the State began to play an enlarged part in rural development on the coast, this new rôle owed much more to petroleum revenue than it did to a change of heart over agrarian reform.

AGRARIAN REFORM IN THE GUAYAS BASIN SINCE 1970

Although the 1964 Agrarian Reform Law devoted only five paragraphs to 'precarious tenancy' on the coast, it was in this region that the most dramatic changes were to come about after 1970.* In December of that year, a Government decree, *Decreto 1001*, was issued which was to acquire increasing significance when foreign exchange earnings from petroleum removed the obstacles to financing an agrarian reform. *Decreto 1001* was one of a series of measures which aimed to abolish 'precarious tenancy' on the coast and stimulate landlords to invest more in their estates. There were a number of new features in *Decreto 1001*, however, which distinguished it from other agrarian legislation in Ecuador. This decree provided the Land Reform Institute (IERAC) with greater powers of

* The term 'precarious tenancy' describes a variety of kinds of tenancy, in which cash played no part, and in which the tenant was subject to considerable exploitation. Both the *'huasipungo'* system in the sierra and the rice *'precaristas'* on the coast were 'precarious' tenants.

intervention than it had previously possessed, and envisaged the setting-up of marketing and production cooperatives composed of former tenants. IERAC was able to intervene on any estate where it suspected that landlords were employing tenants. In the course of time, land titles would be issued to these former tenants, which would enable them to obtain agricultural credit from the Development Bank. Through its possession of both a 'carrot' and a 'stick', the Government would be able to dictate the form that development took in the 'reformed sector' of coastal agriculture. At the same time, it could guarantee supplies of a staple foodstuff, rice, to the urban market.

Many of the provisions of *Decreto 1001* were later incorporated in the 1973 Agrarian Reform Law, which placed an even stronger emphasis on technically advanced, and commercially viable, farming. Public funds were increasingly deployed in these new 'agricultural enterprises', building up the physical and social infrastructure which had been lacking on most estates. Only the most economically successful 'cooperatives' were given this degree of attention, benefiting from what the first President of the 1972 Military Government termed the 'sowing of oil' (*la siembra del petróleo*).

In some respects, the rice zone of the Guayas basin, comprising the provinces of Los Rios and Guayas, provided an ideal region for Government intervention. Rice was produced on estates of varying size, very few of which employed advanced technology, irrigation or chemical fertilizers, and pesticides. Even some of the largest and most modern estates also employed tenants who worked the land on a one-year leasehold arrangement, entered into orally with the landlord. These tenants were descended from a number of different racial groups which had settled on the coast, including Asian indians, arabs, negroes, and Chinese. Periodic seasonal migration from the Andean provinces had also influenced the ethnic make-up of the population. When the cocoa estates had fallen into decline, many of the tenants who worked them began to cultivate more rice on the flooded land within the estates. In other cases, the banking sector bought up estates from their bankrupt owners, and encouraged rice cultivation. The tenancy system that evolved was thus a recent development, with clear advantages to landlords, many of whom were absent from their estates for most of the year. It was not an archaic 'feudal' institution, like the *huasipungo* system in the sierra.

Landlords in the rice zone were intimately involved in the rice economy even when they preferred to live in Guayaquil, or one of the other coastal towns. The landlords owned the rice-mills, which bought the tenants' rice at depressed prices, and frequently sold it again at great profit. By controlling the supply of processed rice, the millowners were able to exploit production shortages, and were able to maintain prices during good seasons by stockpiling supplies. Many landowners and millowners also acted as moneylenders to the tenants, making credit available to meet regular subsistence needs at usurious rates of interest. It was calculated that millowners were able to pay for the capital cost of building and equipping a small mill in the course of three years of market dealing.

There were a number of advantages to be derived from employing rice tenants in place of wagelabourers. Most estates were seriously undercapitalized, tenants employed only a few simple tools in the production process, and rice was only cultivated on land that had been flooded. Tenants worked much longer hours than wage labourers for the same income, and employed labour from within the family, meeting many of their subsistence needs themselves. Labour was in plentiful supply, and land distribution sufficiently 'skewed' to allow landlords to exact a considerable surplus from their tenants. This surplus took the form of a transfer of value derived from the exploitation of cheap labour, rather than the payment of extortionate rents. It was the landlords' and millowners' control of the commercialization process (the processing and marketing of rice) that enabled them to make substantial profits without improving the technical foundations of production.

After a series of disastrous droughts and crop failures, which led to the Government importing large quantities of rice, pressure began to grow for a radical reform of the production, processing, and marketing of rice in Ecuador. The Guayas basin, where rainfall during the first three months of the year was of 'monsoon' proportions, held the promise of enormous production gains if land was properly levelled and irrigated. Increasingly, technical experts, especially foreign advisors, urged the Government to abolish the tenancy system and rationalize the agrarian structure of the zone (CEDEGE, 1970). By 1972, as we have seen, the necessary finance was available; the principal obstacle to development had been removed.

Another important factor which persuaded Velasco Ibarra's Government to abolish rice tenancy was the agitation mounted by the tenants themselves. Such was the scale and intensity of this agitation during 1968 and 1969 that it became impossible to sell land in the rice zone. The unwillingness of prospective landlords to buy land, the unwillingness of existing landlords to modernize their estates, and the pressure mounted by tenants for land all contrived to 'freeze' the land market (Zuvekas 1976). The discontents of the tenants were given encouragement by the Agency for International Development (AID), which was active in the area, helping to establish rice cooperatives and make legal advice available to tenants. AID staff in Ecuador insisted that an amicable arrangement could be made between landlords and tenants, under which the ownership of land would be transferred to cooperatives, by means of financial guarantees from the banking sector.

This plan, originally called 'Land Sale Guaranty', was taken over by the Military Government after 1972, but many of the ideas that inspired it were retained. Henceforth, the State, rather than a North American agency, took responsibility for organizing former tenants into cooperatives, on terms established by the Government bureaucracy. Agrarian reform in the rice zone has done something to meet the demands of tenants for land, but the main goal of the reform has been to increase production. Furthermore, the origins of the reform, as we have seen, lie in private initiatives to allay discontent and guarantee food supplies, rather than radical political objectives.. The cooperatives that have received most Government support are in a minority. In most cases, technical and management decisions on these cooperatives are made by personnel employed by the State, rather than the cooperatives' members. Increased capital investment, whatever improvements it has brought in production terms, has also served to exacerbate the employment situation. Large numbers of coastal labourers can only find irregular seasonal employment; many more combine wage employment with the sale of small production surpluses on the market.

The new agrarian structure in the rice zone shows signs of producing new sources of tension between cooperatives and within them, rather than between landlords and tenants. Former tenants are organized in several 'federations' of rice cooperatives, each seeking to obtain maximum assistance from the State, and reflecting the various political objectives of their constituent parts. Differences within cooperatives and between cooperatives are thus reflected in contradictory demands, and an inability to make common cause in their relations with Government and landlords. In general terms, one might argue that the problems of absentee and exploitative landholding and marketing arrangements have been replaced by those of capitalist agriculture, where labour is shed, rather than employed productively throughout the year, and political 'deals' with the State have replaced landlord patronage. For those peasants within the new cooperatives whose livelihoods have improved, there are new problems: how to ensure that their children find employment, that their leaders are accountable to them, and that collective expenditure does not lead to indebtedness and greater dependence. Only a more systematic and global agrarian policy, which establishes who is to benefit from the terms under which the agricultural surplus is extracted, can begin to solve the structural problems of coastal producers in the future.

SINGULARITIES OF THE LAND REFORM PROCESS IN THE SIERRA

While the process of land distribution and the creation of new farms has proceeded most rapidly in the coast and eastern jungle, the highlands have been faced with the necessity of a different sort of land reform. The most glaring inequalities in the distribution of land and the most feudal forms of land tenure have been largely confined to the highlands. If land reform was to transform the structure of rural society, it would have to do most here, notwithstanding the changes that would need to take place on the coast. The singularity of the changes since 1964 in the landholding pattern in the sierra lies in the impact of nominal land redistribution on many of the estates, and in the equally nominal expropriation of a series of large estates. The land redistribution, as we shall see, was nominal because it involved taking away plots from estate workers with one hand and giving them other plots with the other. The expropriation of estates was nominal because it involved only a tiny fraction of the total area in *latifundia*.

The ending of feudal forms of tenure, of the institution of the *huasipungo*, and other similar forms of labour service, affected the great majority of the highland estates but only a very small proportion of the total agricultural area (3.8 per cent), and only about 6 per cent of the rural families with access to land.* It is impossible to know to what extent all such forms of tenure have been eliminated, but IERAC have long asserted that the distribution of land to *huasipungueros* has been completed. The total number of titles issued is not far short of the number of such holdings in the 1950s. In fact, many estate owners succeeded in selling this land to their former serfs without the intervention of the State.

The formal expropriation of estates owned by large landowners may be viewed in a more realistic perspective when it is realized that over half of the properties affected, and 42 per cent of the land redistributed in the sierra by 1973, belonged, in fact, to either the Government or the Church. A series of large estates, formerly owned by the Church but confiscated in the early part of the Republican period, were administered by the National Social Welfare Group (Junta de Asistencia Social) and the proceeds were used to support hospitals in regional capitals, among other things. The estates were run in much the same way as other large properties, and in need of as much reform as any other large estate. The Church properties, one of which was owned by the Diocese of Riobamba (whose bishop is a noted left-wing figure), were equally backward, but were made readily available for expropriation. The Government estates were the subject of numerous expensive studies to establish how they could best be subdivided and were used as showcases, indicating what could happen with a Government-led land reform. That such former estates are now less often used as examples of the benefits of a benevolent agrarian reform is testimony to their failure to impress visitors and their patently atypical nature.

A further singularity of the land reform in the sierra is the extent to which it has been an additional factor in stimulating a part of the rural population to migrate elsewhere, either to rural areas where there is land and employment more readily available, or to the towns and cities. This migration is associated with land tenure changes because of the decline in the availability of land to rent, and the inadequacy of the size and quality of the plots of land received from former landlords.

SPECIFIC CHANGES IN THE SIERRA DURING THE REFORM PERIOD

Many of the changes in farming and land tenure that took place in highland Ecuador from the early 1960s should be seen as associated with the threat of land

---

* Calculated from agrarian reform data for 1964–73 and provisional results of the 1974 agricultural census.

reform quite as much as the actual provisions of the Agrarian Reform Laws of 1964 and 1973. The following examples of the sorts of changes which have occurred in several areas of the highlands are drawn from a recent study of rural emigration and agricultural development in this area, in which detailed analyses were made of changes in five case-study areas (Preston et al., 1978).

The major changes in rural areas following the agrarian reform laws may be grouped under three headings: the decline in use of tenancies by large landowners, the redistribution of *huasipungo* land, and the sale of land by the estate owners which further affected redistribution.

The main reason for the decline in the importance of renting for cash or for a share of the harvest was a fear of the laws being implemented in favour of sitting tenants. The laws provided that if a tenant occupies the same area of land for more than three to five years, then he may have the right to acquire it from the landowner. The provisions of the laws have not been widely implemented, and the exact provisions of the rental contract that might bring it within the definition of affectable property are uncertain. Many landowners who had previously rented land, frequently for a share of the crop, often negotiated different contracts with each individual. The fluctuating attitudes of successive Ecuadorian Governments towards agrarian reform caused many landowners to do what they could to minimize the effect on them of a reform, if it was ever implemented. Consequently, in many cases, landowners ceased to lease their land, but instead worked it directly themselves, using locally recruited labour. In other cases, the farming system was changed to enable the land to be farmed with a minimum of direct labour, since the cost of labour was now rising with inflation and an increased awareness of the wage rates that prevailed in towns and colonization areas.

A parish in Pichincha province may serve as an example of these changes. Nono lies in a well-watered valley to the west of Quito, where a series of estates contained most of the best agricultural land, and the mestizo population of the village was mainly made up of sharecroppers and day labourers, plus a few *huasipungueros*. During the 1960s, the estates gradually reduced the labour force, switched from arable farming to dairying, ceased to offer land to sharecroppers, and began employing women from the village as milkmaids. The number of *huasipungueros* was reduced, both in Nono and in the parish to Atahualpa to the north, where many peasants reported consistent ill-treatment and the abuse of serfs and sharecroppers. This abuse eventually forced the tenants to leave the estates voluntarily, as had been intended by the landlords, thereby reducing the number of families that might eventually be entitled to claim land of their own from the estate. In other cases, such as Hacienda Ayacón, in an indian area near Guamote in Chimborazo, thirteen serf families were expelled from the estate in the early 1960s and were accommodated by the adjacent freeholding community of Santa Teresita.

Data from the rather inadequate agricultural censuses of 1954 and 1968 indicate the extent to which the increase in rented land units was less than in other categories, in part as a result of the decrease in tenancies on estates. While in the coastal region the extension of land worked by tenants had increased between 1954 and 1968 by 237 per cent, in the highlands only a minimal increase (0.3 per cent) is noted. In Azuay province, the number of tenants actually decreased over this period, although the area of land farmed by them increased slightly.

The distribution of plots of land to former 'serfs' was completed in all the estates visited by Preston and Taveras, although in a number of cases the land had actually been bought by the estate workers, who anticipated a long-drawn-out and less-favourable settlement if the Agrarian Reform Institute was involved (Preston et al., 1978). The extent to which former 'serfs' had received land is indicated in data from 194 of the interviews with former estate 'serfs' 71 per cent of whom had not received any land from the landowner. Considerable variation occurred between the study in Loja province, where 54 per cent of workers had received land, the Pichincha study, where 32 per cent received land, and the three other studies, where less than 9 per cent of the workers had received land (Taveras, 1977). Even making allowance for workers occasional exaggerating their status to that of *huasipunguero*, this suggests that a major part of the estate labour force was dismissed without receiving any land. Many of the beneficiaries received about the same area of land that they had previously farmed (39 per cent of the sample), although almost half of the farmers (48 per cent) were now worse off than they were before receiving land (Taveras, 1977, p. 18).

The most important change of all in the post-1963 period has been the large-scale sale of land by estate owners, either selling up their entire holdings or selling the majority of their land to avoid expropriation should the existing land reform laws be applied. In addition, an increasing number of the ageing highland estate owners and their heirs no longer wish to be clearly identifiable as land monopolists, i.e. targets for political demonstrations, and even for criticism from radical priests. The dramatic increase in the number of holdings shown by the 1954 and 1968 census data reflects the tendency to sell off land, as well as the break-up of myriads of farms by inheritance. No data can adequately document this process. However, it is worth noting that on only one estate in all those studied by Preston and Taveras was land sold exclusively to former serfs. Although sales of land have increased, particularly in the 1960s and 1950s in the indian area of Guamote, Chimborazo province, sales of appreciable quantities of land to groups of indians have been made since the early part of this century.

Several groups of people have been able to benefit from these sales. In some cases, former 'serfs' have been able to buy land in addition to that allocated them in lieu of a *huasipungo*, if they can mobilize sufficient capital. Further, small-scale merchants in parish centres and middle-sized farmers with access to capital have bought either cultivable land on which they have built a home (as in some areas of Loja province) or pasture land at high altitudes, where livestock can be grazed. The number of people who have acquired land in this way was small in the study areas cited above, and a number of these people live in other parishes. Finally groups of peasants have also bought land in this way, either for members of a community or a group formed into a legally recognized cooperative.

AGRARIAN REFORM AND EMIGRATION IN THE SIERRA

The redistribution of land effected directly and indirectly as a result of the land reform laws seems to have benefited both the better-off 'middle' peasants and village capitalists, but it has also introduced an element of inflexibility into the tenure system. Martinez Alier has already indicated the relatively privileged status of estate workers in Peru (Martinez Alier, 1977) and 40 per cent of the former *huasipunguero* informants interviewed said that it used to be easy to obtain an extra plot for their children (Taveras, 1977, p. 11). Once land had been allocated to them by the estate, most ties with the estate were severed and succeeding generations of children had to subsist on an even more inadequate holding. The alternative was usually to migrate to find work elsewhere, at least for part of the year.

The former 'serf' was often himself forced to migrate when the production from the new land proved inadequate to support his family, and in any case he often had inadequate land to allow a crop rotation that would preserve soil fertility. Formerly, *huasipungos* were often moved as the land was exhausted; now this was impossible. The former serfs have joined the landless in their migration, those who had previously worked on the estate either as tenants or in some other menial capacity, but for whom there was now no further employment or land. Finally, the dispossessed 'serfs', banished from land farmed for generations, were left to fend for themselves. Those like the workers on Hacienda Ayacón, already mentioned, who were received by a community or acquired land by some other means, were in a minority. Many others were forced to move elsewhere to the towns or to areas of new land settlement.

AGRARIAN DEVELOPMENT AND AGRARIAN REFORM

In conclusion, it is worth making some observations on the nature of recent agrarian development in Ecuador, and its relationship to agrarian reform policy. Although the agrarian structure of both major regions is so unalike, what emerges is the similarity of social processes in both regions and the complementarity of agrarian policy and spontaneous development processes.

Both coast and sierra have witnessed the abolition of 'precarious' forms of tenancy, and the increased use of wage labour, both on private estates and within the coastal 'reformed sector'. The beneficiaries of this process have included those landlords who have been able to sell off land, in anticipation of the law, and those 'middle peasants' who possessed enough capital to buy land. In both regions, the legal obligation to transfer land to former tenants has not prevented private land sales from being effected. It has been suggested in this chapter that the ideological inspiration for the land reform legislation was essentially liberal, and that legislation sought to reinforce, rather than to challenge, existing processes of social differentiation and the 'proletarianization' of the rural labour force. In neither region has the increased rural population been absorbed into employment on the land. Migration to the cities has increased. Even the opportunity for dramatic increases in agricultural production in the coastal rice zone, which the 'windfall' from oil provided, has done little to improve the living standards of either the rural or the urban masses.

BIBLIOGRAPHY

CEDEGE (Comisión de Estudios para el Desarrollo de la Cuenca del Rio Guayas), 1970, *Tenencia de la Tierra y Reforma Agraria—un Estudio Socio-Económico y Legal*, (Guayaquil: T. Ingledow and Associates Ltd./Guayasconsult).

CIDA (Comite Interamericano de Desarrollo Agrícola), 1965, *Ecuador: Tenencia de la Tierra y Desarrollo Socio-Económico del Sector Agrícola* (Washington, D. C.: Unión Panamericana).

Martinez-Alier, J., 1977, *Haciendas, Plantations and Collective Farms* (London: Cass).

Preston, D. A., Taveras, G. A., and Preston, R. A., 1978, *Rural Emigration and Agricultural Development in Highland Ecuador*, Final Report to U.K. Ministry of Overseas Development, Working Paper 238, School of Geography, University of Leeds.

Redclift, M. R., 1978, *Agrarian Reform and Peasant Organisation on the Ecuadorian Coast*, Monograph 8, Institute of Latin American Studies (London: The Athlone Press).

Taveras, G. A., 1977, *Tenencia de la Tierra, Reforma Agraria y Migración en la Sierra Ecuatoriana*, Working Paper 201s, School of Geography, University of Leeds.

Zuvekas, C., 1976, 'Agrarian reform in Ecuador's Guayas River basin', *Land Economics*, **52**, No. 3, August.

Colonization

Environment, Society, and Rural Change in Latin America
Edited by D. A. Preston
© 1980 John Wiley & Sons Ltd.

CHAPTER 5

From colonization to agricultural development: the case of coastal Ecuador

ANNE COLLIN DELAVAUD

1. INTRODUCTION

Western Ecuador is a vast region of some 70 000 km^2 which today demonstrates such a variety of methods of development and of ways of organizing farming that it offers a valuable range of experience from which to study the transition from extensive to intensive farming. The *Costa* of Ecuador owes its variety of experience not only to the diversity in its natural environment but also to the demands of foreign markets which, in less than a century and especially in the last thirty years, have brought about considerable changes in settlement, productive systems, and regional development.

If the enormous forested areas of the northern *Costa* are made available for spontaneous colonization by small-scale farmers, and everywhere else is given over to large estates, the areas of small farms, such as the old colonization areas, will have to respond to the demand for foodstuffs in order to provide a living for their workers. The contemporary modifications in farming are associated with changes in tenure and with new methods of organizing work that are also associated with differences in cultural relationships and in use of labour.

2. EXPORTS AS A DRIVING FORCE FOR COLONIZATION

2.1 Cacao and early development

The economic history of the *Costa* is one of a series of changes in the agricultural economy imposed by external market constraints, organized by exporters and large landowners, in order to overcome the obstacles presented by the physical environment. The rise of the agriculture of the coastal region is closely associated with its proximity to the Pacific coast, which facilitates access to overseas markets that are the envy of the colonists in the *Oriente*, to the east of the Andes, whose access to foreign markets is difficult and costly. In addition, the commercial dynamism of Guayaquil attracts, to the coastal lowlands, workers from coastal Manabí province and from nearby indian areas in the highlands. The establishment of cacao plantations 150 years ago in the Guayas basin marked the beginning of commercial development of a region whose agriculture had previously been confined to the levées of the Daule river where food crops and tobacco were grown. In 1918, almost 200 000 hectares of cacao were established along the Vinces and Babahoyo rivers, which enabled easy transport of the cacao beans to the port of Guayaquil. Small cacao zones grew up in the extreme south of Guayas province, in El Oro, near Chone in Manabí, and near to the town of Esmeraldas. Ecuador was then the largest exporter of cacao in the world.

After several decades of prosperity and growth, there followed a long crisis as a result of falling world prices, competition from new cacao-producing countries, and also plant diseases. This was the first check to the progress of colonization: the cultivation of subsistence crops and rice near to abandoned cacao plantations enabled the workers to eke out a living while awaiting better days. Coffee and rice became more important, but they were not as profitable as cacao, and this period saw the first exodus of migrants to the towns, and in particular to Guayaquil. Only sugar-cane, established at the end of the nineteenth century, grew in importance, particularly on the large estates near to Milagro.

2.2 The advance of the agriculture areas and the banana boom

The period 1940–45 caused hopes to rise with the strong demand for produce from the countries engaged in the Second World War, in particular for rice, rubber, and balsa wood. This demand continued in following years as international demand for cacao, coffee, and also bananas grew. The rise in importance of bananas was such that the land available in the old cacao areas was insufficient and new lands were necessary. A new pioneer fringe developed near the hills of Quevedo in the northern Guayas basin, in which both small- and large-scale farmers from *Costa* and *Sierra* Ecuador took part. The high wages of the lowlands attracted large numbers of highland indians. Similar changes took place in the Jubones valley in southern El Oro province, which became particularly involved in banana production. Despite its low rainfall and occasional floods, an impressive infrastructure of a port (Puerto Bolívar), canals, and roads developed. In the north, foreign companies attempted to repeat the experience of the Guayas basin by clearing the banks of the Esmeraldas river, but they suffered from the lack of either a hinterland or an active commercial bourgeoisie. Only coastal Manabí, the arid Santa Elena peninsula, and the jungle of northern Esmeraldas were little affected by the banana boom, but these areas did contribute migrants to the growing labour force on the new plantations, although many other migrants came from the Andes provinces.

Starting in 1947, in less than seven years 153 000 hectares were producing bananas for export, and this area was maintained until 1964 without a major crisis. The soils were fertile and there was abundant land available. The agricultural frontier advanced northwards in response to growing demand, and tracks gave way to roads which replaced river transport and the needs for transshipment. The banana boom was suddenly checked by the introduction of a new variety of banana following increasing losses as a result of Panama Disease—the Cavendish variety replaced the traditional Gros Michel. The superior yields of the Cavendish and increasing competition on world markets associated with other countries increasing production brought about a reduction in the area devoted to commercial banana production. The areas with most level land and best access to the major port were those where banana production continued to flourish—particularly, therefore, in the provinces of Guayas and El Oro, rather than the new pioneer areas further north.

2.3 Continuing colonization

The banana boom attracted colonists from every part of the country, and from a variety of social backgrounds. Both farmers and professional people were ready to plunge into the forests near Quevedo where the State readily gave them plots of 50 hectares to increase the area growing bananas. In less than three years, the costs of installation and of clearing the land were covered, since banana cultivation was both profitable and straightforward. Starting with only limited means, the colonist could become an owner–producer of bananas for export in a short time. The large landowners of Guayas also profited from the demand for bananas and extended the areas in bananas on their land. And, in particular, they bought properties in the new pioneer zones as new highways were constructed allowing easy access to Guayaquil port.

The State was involved in colonization in the Santo Domingo area, where a small number of estates existed on a huge alluvial cone at the mouth of an Andean valley near Quito. Lots of 50 hectares were distributed to colonists, most of whom were townsfolk, to grow cash crops. However, even before the titles of the plots had been issued and the provision of an infrastructure completed, the banana boom, which had had a great impact on this region, was over. The area was far from a port, and the land was sloping and not fully suited to the planting of the Cavendish fruit. Farmer's cooperatives, which were anyway suffering from an inadequate financial and technical support, found themselves against the wall. For the former townsfolk, this meant a retreat to their urban points of departure. But, for the other farmers, it threatened penury, since the Government had little idea of what sort of agricultural programme should be followed in the new colonization areas, where many new farmers had cleared land on the edges of the area of planned colonization. Today, contrasts still exist between the areas of spontaneous and planned colonization, where the former areas struggle to adapt to the new situation, and the latter stagnate while waiting for Government help. This failure had its effect in Government colonization policy, for no subsequent

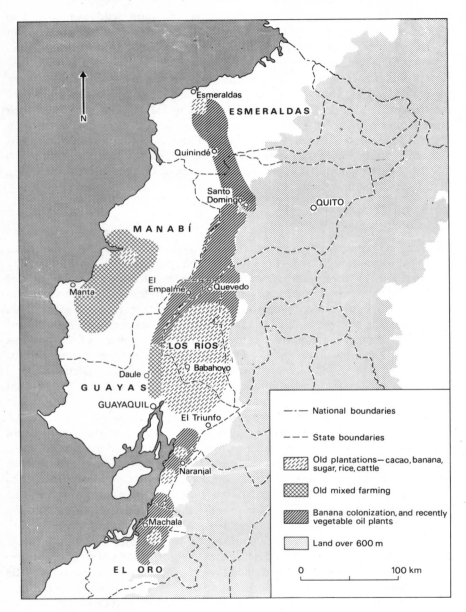

Figure 5.1 Agricultural development in coastal Ecuador

colonization project has been undertaken on the coastal lowlands. Likewise, no private land development companies, selling land and initial assistance to colonists, exist any longer.

Spontaneous colonization does continue largely on land owned by the State, where settlers, often without title, grow food crops and a few cash crops dependent on the level of isolation from markets. Although less well known, these migrations have led to the settlement of large areas, especially the hills of interior Manabí province and the slopes leading down to the Guayas basin. This same migration from Manabí has even spread settlement over into the Esmeraldas basin. On a lesser scale, migrants from the sierra have settled in the foothills of the Andes and in the abandoned areas in the older farming zone, in particular in southern Guayas and in El Oro province.

3. COLONIZATION AND THE TRANSFORMATION OF THE COAST

3.1 The extension of the cultivated area

The coast has changed more in the past thirty-five years than in the preceding two centuries. The banana boom led to the clearing of more forest and such an advance of the agricultural frontier that, once the apogee of development was passed, the cultivated area was too large. The recent stimulus to export has thus led to the recultivation of abandoned land and an intensification in land use, as well as renewed clearing of new land by small-scale colonists. Figure 5.1 shows agricultural development in coastal Equador.

In twenty years, some 730 000 hectares have been cleared for farming in the coastal provinces, which is an apt measure of their abundant land resources.

Table 5.1 Agricultural change on the coast 1954–74[a]

	Number of holdings			Area in farms (1000 ha)		
	1954	1974	Change (%)	1954	1974	Change (%)
Esmeraldas	6 677	14 832	122	171	519	203
Manabí	38 028	64 397	69	982	1 274	20
Guayas	22 831	47 641	108	1 023	1 052	0.2
Los Rios	8 779	28 303	222	600	561	−6
El Oro	8 400	14 077	67	199	299	50
Total	84 715	169 250	99	2 975	3 704	24

Source: Censo Agropecuario, 1954 and 1974.
[a] Some reservations exist about the comparability of these data in the two censuses as a result of a change in the definition of land holding.

These changes have not affected all provinces equally, and in Guayas and southern Los Rios the cultivated area has scarcely increased from 1954 to 1974 (see Table 5.1), but a more effective use of the resources has been developed as a result of the increase in number of holdings, and the subdivision of holdings, sometimes under threat of agrarian reform. Los Rios, if the 1954 Censo Agropecuario data are reliable, has actually experienced a decline in its cultivated area during the 1954–74 period, although the number of holdings has increased dramatically (222 per cent). In El Oro and Manabí, however, where both the physical environment and the manner of development have been different, the cultivated area has increased by 50 and 20 per cent, respectively, and the number of holdings by 67 and 69 per cent. Finally, Esmeraldas emerges as the main colonization area of the last twenty years, with the cultivated area tripling and the number of holdings more than doubling. The increase in area cleared for cultivation has also affected the western part of Pichincha province, in particular along roads radiating from Santo Domingo de los Colorados.

3.2 The increasing demographic importance of the coast

Increases in the population over the last thirty years have made the *Costa* the most populous region of Ecuador, having now overtaken the *Sierra*. A high birthrate and a positive balance of migrants, especially from the sierra towards the coast, has enabled the population on the coastal lowlands to increase from 1.3 million in 1950 to 3.2 million in 1974.

The movement of colonists has mingled people from various regions: indians and mestizos from Manabí and the mountains occur in the Guayas basin and are now spreading into the southern part of the area occupied by the negroes. Despite the general distribution of population throughout the coast, considerable regional differences exist, and this semi-arid zone, in particular, has very high population densities. In Manabí, in spite of emigration, there is still overpopulation in relation to available resources. Guayas remains the most populated of the coastal provinces, and contains half of all the *Costa* population. Half of these live in Guayaquil, which continues to attract migrants whatever the possibilities of lodging and employment. The estates of Guayas, after having absorbed their necessary labour force, see the remainder go to the small towns and to the port of Guayaquil, which also attracts thousands of migrants annually from the highlands.

The movement of colonists to the north and south-east is also associated with the creation of new urban centres at the junction of roads or rivers such as Quevedo, Santo Domingo, Machala, Quinindé, El Empalme, El Triunfo, and Naranjal. The first three towns grew up during the banana boom and soon grew in importance as regional centres at the expense of the centres of the cacao boom, which were nearer to Guayaquil. Urban growth is widespread in the *Costa*, and the number of urban centres with more than 2000 inhabitants rose from twenty-four in 1954 to seventy in 1974. The urban network is dominated by Guayaquil, which has a population of 850 000, while the next two urban centres in order of size (Esmeraldas and Manta) had less than 70 000 people in 1974.

4. NEW ORGANIZATION OF COASTAL AGRICULTURE

4.1 The improvement of exports and the growth of internal demand

Petroleum resources have given the Ecuadorian State far-superior financial backing than was previously available. Although industry and commerce have benefited from petroleum revenues, agriculture has also been able to benefit. In 1975, the agricultural sector was declared a priority area for financial investment in order to try to overcome the decade and a half of worrying stagnation that had affected the main agricultural areas. Agricultural production had been growing less rapidly than national population, and even the export sector has suffered from international competition and financial difficulties which have hindered the adaptation of estates to the needs of foreign markets.

After a series of difficult years, the export-oriented estates benefited from a general rise in export demand for a whole range of tropical products in the early 1970s, which stimulated improvements and specialization in new crops. Castor oil, soya, ground nuts, oil palms, hemp, and cotton all now have a more important rôle in coastal agriculture. The traditional agricultural exports of cacao, coffee, and bananas are still important and, in 1977, accounted for 62 per cent of exports other than petroleum, worth $ 350 million. But, in the period since 1972, there is further evidence of diversification, with the rise in importance of fishing and of industrialization, and of a gradual increase in exports such as processed cocoa, sea food, fibres, and domestic electrical goods, as well as petroleum (see Table 5.2).

Table 5.2 Ecuador's principal exports ($m FOB)

	1970	1972	1974	1976	1977
Bananas	83	109	113	136	131
Cacao beans	22	23	102	32	56
Coffee	50	46	67	205	148
Sugar	8	17	44	6	14
Processed cocoa	2	6	22	62	58
Fish products	4	18	28	54	61
Petroleum	0.8	59	526	565	461
Total	189	300	962	1 127	1 120

Source: Anuario de Comercio Exterior, 1978.

The sources of agricultural exports are more than just the Guayas basin and Guayaquil, and include all of the coastal zone from which come fish products, fibres, and vegetable oils. Finally, domestic demand for foodstuffs for the growing urban areas and for raw materials for industry throughout the country has stimulated agricultural production from small and large properties. This commercial revival, which was hampered by bad weather affecting farming, eventually got under way in 1975. It was necessary because, as petroleum exports grew in importance (for example, in 1973), there was a general drop in agricultural production, steep rises in prices, and it was even necessary to import cereals.

4.2 State intervention in agricultural modernization

Agricultural changes have been associated with major assistance from the State that has modernized the production of export crops, and the diversification of cash crops and the development of spontaneous colonization in Esmeraldas. Whilst the Government has encouraged agriculture in the sierra, it has recognized the greater agricultural potential of the coast which can be realized by means of modernizing the methods of production and by public works projects. The new Agrarian Reform Law of 1973 indicated the Government's new attitude, giving priority more to improved land use than land redistribution.

The creation of Development Programme intended to encourage one or sometimes several agricultural products has introduced an important link between producers and merchants. Each Programme is given a budget by the Ministry of Agriculture with which it promotes the production of the particular crop in the areas of the country that are most appropriate to it. In addition, the National Institute for Agricultural Research (INIAP) develops and offers for sale the best varieties of seeds for the areas in which production is being developed. The Programmes offer the farmer a team of technicians with relevant information and lines of credit through the National Development Bank which facilitate the improvement or increase in area of the relevant crop. Credit proposed by the Cotton Programme in 1972–73 was over $80 000 and rose eightfold in the following crop year. These new possibilities for advice and credit have also been taken advantage of by the larger estates and, although it is too soon to know the latest regional breakdown of credit distributed by the banks since 1975, at a national level credit authorized to the agricultural sector rose from $40 000 in 1971, to $192 000 in 1974, and to $324 000 in 1976, thus almost doubling in the last two years for which data are available. The better access to credit has made more money available to small-scale farmers. It is no longer necessary to have a land title to borrow money, and formal recognition of occupation by the Agrarian Reform and Colonization Institute (IERAC) will suffice. This opens the way to credit to a large number of farmers who have to wait many years to obtain legal title to their property and who have previously only been able to borrow money from local moneylenders and merchants who charge a high rate of interest,

The Government has further aided agricultural development by easing import restrictions to allow the importation of seeds, agricultural machinery, or fertilizer and by planning the construction of storage silos on the coast to improve the inadequate existing storage facilities. These new measures have not been without their share of problems, because the stimulation of a particular crop causes competition between seed dealers, merchants, transporters, and processing industries, so that there is often a sharp difference between what the farmers produce and what the market can cope with. In some cases, crops being pushed by the Government have not found a ready market.

4.3 Infrastructural improvements

The growth of commerce stimulated by increased colonization in the *Costa* has only been made possible by the creation of a new infrastructure. While rivers and mule tracks have continued to provide access to some long-settled areas, the construction of highways has aided the development of estates oriented towards the production of commercial crops for export or for industry. Constructed in great haste in response to the banana boom, the road network now favours banana zones and access to the port of Guayaquil, at the expense of east–west highways and the backward regions. The coast now has some 3120 km of paved roads, compared with only 1400 km in 1965, and 51 000 vehicles were registered in 1975, compared with 17 800 registered in 1965. The north and south frontiers from Esmeraldas (and soon Limones) to Huaquillas are linked by a 765 km highway. These road improvements are related also to the creation or improvement of inter-city highways to the coast and of links with the sierra. The Government has also built roads and cleared tracks in Esmeraldas to help pioneer farmers establish themselves, as was the case near Quevedo and Santo Domingo and, more recently, in the *Oriente*. Port-works, begun in the previous decade, are nearing completion. A new motorway is planned which will link Quito with Guayaquil in one direction through Quevedo, and in the other further east through the towns of the central sierra. The Guayas basin is to be the centre of major irrigation and drainage works supported both by Quito and Guayaquil, and will perhaps eventually be a major centre of agriculture, industry, and capital.

4.4 Development projects

The carrying-out of engineering work to provide irrigation as well as protection against flood control is the best way of improving agriculture in many areas of the Ecuadorian *Costa*. In the long term, all of the Guayas basin will benefit from the construction of the Daule Peripa dam—a huge reservoir—which will be capable of controlling flow on the Daule river and of lessening the flooding in the central part of the basin. It will also enable the irrigation of many areas, and will even take water as far as the arid Santa Elena peninsula. This project, undertaken by CEDEGE, should put an end to the alternating disasters of drought and flood which at times destroys as much as 30 per cent of the harvests, and often even more for small-scale farmers.

The Babahoyo Pilot Project, whose infrastructure development is almost complete, comprises the development of 11 500 hectares of irrigable land. Strong incentives to plant more rice are provided by offers of seeds, fertilizers, and also heavy machinery for the construction of canals and access roads. Some sixteen cooperatives include families that have received land from former estates through expropriation, and they work collectively. Southern Guayas likewise benefits from construction projects, such as the Milagro Project which will enable the irrigation of 5000 then 10000 hectares. In Manabí, the canals from the Poza Honda dam are soon to be completed, as well as the Carrizal–Chone canal. Finally, in El Oro, the Jubones project will improve farming in an already developed area.

These works have largely been financed by international agencies, in particular their initial stages of investigation, but since 1975 banks have been required to invest 20 per cent of their commercial portfolio in national agricultural development, either by buying special bonds carrying 5 per cent interest or by direct investment. This has succeeded in releasing even more Government funds for further necessary investment.

4.5 The continuation of spontaneous colonization

The recent developments to which we have previously referred have re-emphasized the national importance of coastal agriculture, and the trend towards more rational land development has been noted, particularly in the Guayas basin. This does little, however, to combat the social problems that are posed by the lack of land and the gradual migration of agricultural workers towards the towns, as a result, in part, of increasing mechanization. The policy of land redistribution envisaged by the latest agrarian reform measures conflicts with the aims of modernization, for it leads to the creation of very small holdings. Furthermore, the law is not applicable to estates other than those where only a small proportion of the land is farmed. Finally, the Tenancy Law of 1970 is only applied in certain areas, especially in the rice zone of southern Guayas, where strong social pressures exist, as described by Redclift and Preston elsewhere in this volume.

Spontaneous colonization remains the only solution for would-be farmers in areas of land shortage or social conflict. Such colonization also enables the initial integration of new lands into the effective national economic territory and a subsequent increase in regional agricultural production. This form of settlement is aided by the prevalence of State ownership of unused forest land, but the subsequent recognition and legalization of ownership by IERAC is a slow process.

Colonists establish themselves wherever they can find free land for clearing as near as possible to a means of access, be it road or river, to the outside world. Heavy rainfall in the northern pioneer fringe aids farming initially, but the one main constraint to increasing settlement is the existence of large timber concessions given by the State to private companies who do all they can to avoid colonists squatting on their land, whatever the size of their concession and even if it has been little used. Land reserved by the State in Esmeraldas has been used by some landowners, in particular near to the Quinindé–Esmeraldas highway, to create large areas of pasture while they await more profitable uses for the land.

The Northern Development Corporation (CORFONOR) is responsible for carrying out an ambitious programme of development in the northern part of the country, including the northern parts of the coast, the highlands, and the eastern lowlands. Among its regional development projects has been the creation, in 1974, of the Esmeraldas Integrated Development Organization (OIE) which, with aid from the Organization of American States, has studied the development of resources in the province and the associated management of capital. The wish to establish more local development administration is a response to the need to study regional problems more closely.

5. CHANGES IN COASTAL AGRICULTURE

5.1 Changes in farming systems

In the past five years, the landscape of the Ecuadorian *Costa* has shown signs of the development of a more intensive farming system. Here and there, in the middle of large traditional estates, large open spaces, levelled, irrigated, and worked by machines, have appeared where previously tree crops, receiving little attention, were grown between drier forested areas and marshes, and which surrounded tiny plots of unirrigated rice land. The main tools for clearing were fire and the *machete*. Now, according to the resources of each farmer, modern equipment may be widely used to shape the new landscape and to permit the development of larger areas than those previously cleared and worked by hand.

The demands of both overseas and national markets are in large measure responsible for the changes that have occurred. Cacao growers, faced with increasing competition on world markets, have had to replant the majority of their old plantations with higher-yielding and disease-resistant varieties. This sort of production is increasingly becoming concentrated in the hands of large estates which specialize in a limited range of crops and can effectively manage marketing to specialized industrial consumers.

The introduction of the Cavendish banana has meant more than just a change in variety—really a complete change in farming system. The Gros Michel fruit served the needs of the colonists of the 1950s, coming from many different areas, but since 1964, as was previously indicated, production has tended to be concentrated in the large estates of Guayas and El Oro, where the expensive investment necessary for packing-stations, machinery, etc., could be undertaken.

Apart from in a few of the commercial farms, coffee has not experienced the modernization drive. Nevertheless, the Coffee Programme has benefited from the huge rise in coffee prices of the late 1970s, and from the buoyant world demand for coffee. Coffee has even become the second most important export crop after petroleum, thanks in part to the major increase in value of the exports. It is doubtful if major changes can easily affect the small-scale coffee grower. Cotton, for its part, suffers from the lack of interest in cotton by large landholders, and it is still grown largely by small-scale farmers who are not encouraged to increase production by the variability of both rainfall and world prices.

Two cereal crops, rice and maize, have only begun to increase in importance as a result of modifications in the cropping methods and the replacement of bananas on some land in the Quevedo area by a rotation of maize with either rice or oil-bearing seeds (e.g. castor oil or soya). Tractors and harvesters, imported at considerable cost, have replaced manual labour to a large extent, while the careful application of fertilizers allows two crops a year to be produced. Most peasant producers of rice or maize still continue to grow these crops alongside cotton or castor oil, as in Manabí, or to have irrigated rice fields. The modest degree of specialization that exists now will, in future, make necessary better use of land resources with regard for climate. In the hills in the north of the Guayas basin, a number of farmers have specialized in oil-seed production, mainly of soya and ground nuts, as well as the long-term planting of African oil palms and hemp fibres. To the east of Guayaquil, major infrastructural improvements have enabled more land to be farmed after drainage or irrigation, sometimes by spraying.

Impressive steps forward have occurred in recent years throughout the *Costa*, but particularly in the Guayas basin, and in the large farms, where one passes from uncultivated or forest land to cleared fields with spraylines aiding the production of two crops a year.

5.2 Changes in the agrarian structure

With each new shift in overseas exports, large-scale farmers and colonists change their farming system. In the first place, the landholding is reduced in order to allow a rational intensification of the cultivated area. The sale of land and successive subdivisions for inheritance have broken up the largest estates that dated from the last century. Recovery after crises which were the result of shifts in demand require larger and larger investments and have necessitated mobility in the land market. The break-up of the estates has aided the growth in the number of small and middle-sized properties bought by colonists, urban people or even given to the farmworkers at times of depression or when the estate owner wanted to get rid of part of his large labour force. The sale of plots of land by landowners in the plantation areas of Los Ríos and Guayas is also a reaction to the possible application of agrarian reform legislation which attempts to find a solution to the occupation of estates by workers to coerce the landowner to sell plots to them.

Despite the general decrease in the size of holding, large landholdings remain in the lowlands, although now these are largely in the hands of private companies, in the hands of Guayaquil financiers or even of overseas interests. Some of these estates have actually increased in size in recent years. In many cases, these were the first properties to have modernized and intensified production for export, while part of the estate was planted to pasture or used for food crops. Two major sugar-cane plantations have kept in production since the last century by regularly modernizing their methods of production, and a new sugar factory is actually now in the process of acquiring land.

The old cacao estates were cultivated by paid workers (*finqueros*), who received a plot to plant in cacao and to maintain, and by *remidores*, who had to give up their plot when the cacao bushes on it were mature and ready for the first harvest, in return for which they could plant food crops between the young bushes. When cacao was no longer profitable, neither the land nor the cacao plantations were considered valuable, and the former paid workers (*finqueros*) stayed on to become tenants who cultivated food crops and especially rice. At first, the workers acquired the cacao trees and subsequently the land itself.

When exports started again in 1940–50, the majority of the former *finqueros* took the opportunity to take over the land they worked on their own initiative, and made themselves into new cacao contractors. Sharecroppers (*desmonteros*) were then brought in to cultivate the rice with little security of tenure. The banana boom accentuated further the contrast between these types of labour on the old cacao estates and those now producing bananas where wage labour predominated. In the pioneer areas, the big estate is surrounded by a reservoir of smallholders (*minifundistas*) used for cheap labour, but elsewhere there is a mass of small and middle-sized farmers anxious themselves to find workers and offering higher wages than elsewhere.

The fear of the application of the Agrarian Reform Law of 1970, and of an active restriction on tenancy and other forms of precarious tenancy, has encouraged estate owners to take over their land again and to take advantage of the new access available to credit funds and to import earth-moving machinery to level fields, plough, and harvest them, as well as to dig canals for drainage and irrigation. Such work needs technicians and mechanics rather than single day labourers with a *machete*. The coming of machinery thus drives away from the countryside the landless labourers, especially in the new areas of cereals and oilseeds which were once areas of major demand for casual labourers. In the areas of the old estates, the slackening in labour demand was slower but still noticeable, and the unwanted workers have left to go to the towns or to the new colonization areas.

5.3 Regional differences in innovations

The change towards intensification is far from being generalized, and the necessary investment is quite unthinkable for the majority of coastal farmers. Such changes, then, are essentially only carried out by farmers with the capital resources to cover the initial costs with the ability to arrange aid from the banks and the Ministry of Agriculture's specific Development Programmes and who can reorganize production in the most appropriate way for their main cash crops. It is particularly the merchants from Guayas, Los Rios, and El Oro who are dynamic and adaptable landowners. Many of them have wide business interests and family connections in commerce and banking, and own several farms. This minimizes the risks involved in commercial farming. The arrival of some Chileans and Peruvians keen for quick profits and with wide experience has further emphasized the general interest in the future of a system of more intensive mechanized farming.

Manabí, too, has been affected by these changes, but they are fewer and less widespread, in part because of severe limitations on commercial farming imposed by aridity, and, there, dry hills alternate with rich and varied farming in the valley floors with fields of 'miracle' rice. Likewise, the Santa Elena peninsula, even more affected by droughts and irregular rainfall, has seen little agricultural change except where water is available. The Santo Domingo area is partially changing, colonization has not really come to an end, and the banana growers have not yet found a better crop. In Esmeraldas, distance from national markets has favoured an increase in livestock rearing by middle-sized and large landowners.

CONCLUSIONS

Colonization has, for a long time, been an important way of seeking a solution to the problem of shortage of land. The increasing population in the semi-arid zone of Manabí, as well as in the Andean basins, has led to a recognition of the need to provide an alternative destination for the migrants who head for the large towns, in particular Guayaquil and Quito. Colonization has thus been directed not only to stimulate agricultural production and commerce but also to settle rural people away from the teeming cities.

The results of thirty years of spontaneous and directed colonization are not entirely negative, even though far more migrants have gone to the cities than to the colonization areas. Despite the wastage of forest resources, many small farmers have established themselves, and a number of larger enterprises have produced export crops. Now the potential is much reduced: the last forests of

Esmeraldas are destined for exploitation for forest products, despite the encroachment of colonists on the periphery; the difficulties of the colonization of the *Oriente* are now sufficiently well known that Government no longer views the Amazon as a major reception area for colonists.

The problem of the numbers of people employed in agriculture, at a national level, has been indicated, and is particularly serious because industrial employment is unlikely to increase sufficiently fast to absorb all the surplus labour. It is, therefore, most important that the intensification of agriculture, already started in the large estates and the areas long-since developed for commercial farming, should be extended to other areas. Much of the intensification which has already taken place has been oriented towards mechanization, which will not provide a solution to problems of rural unemployment. The absence of a Government land tenure policy, even though land reforms of a mild nature have been partially implemented, and the absence of support for stable agricultural prices and the creation of a Prices Board, has not encouraged small farmers to change their farming methods.

Some changes have been noted since the petroleum boom. There has been a considerable supply of credit for agricultural machinery and storage and for the provision of agricultural supplies such as seeds, chemicals, etc., and this does, in part, benefit the small farmer who is a member of a cooperative. In addition, the State has followed a policy of initiating major regional infrastructural projects, but they take a long time to get beyond the planning stage and to affect farmers. Highways, dams, and drainage works are all capable of benefiting the countryside, which lacks such facilities, rather than Guayaquil.

All these efforts are not so much solutions to agricultural problems but rather may provide a basis, hitherto lacking, for the first stage of colonization. The good fortune to be a country which is still underpopulated will perhaps enable Ecuador to develop successfully and to surpass the levels of development of her neighbours. Demographic growth is strong and the growth in internal demand already exceeds the growth of production to the point at which foodstuffs have to be imported just at the time when petroleum should be bringing some prospect of change.

CHAPTER 6

Mexican colonization experience in the humid tropics

JEAN REVEL-MOUROZ

> In my view the future of agricultural production lies in the fertile lands on the coasts. A march to the sea would alleviate the congestion of our central tableland. (Avila Camacho, 1941)
>
> Now let us advance to the sea. Its response will be generous. (Lopez Portillo, 1978)

INTRODUCTION

After four decades and six Presidential terms, 'Advance to the Sea', decreed by President Avila Camacho, still remains an important political aim and is still a fundamental axis in Mexican agricultural development, agrarian reform, and territorial development policies (reviewed in Revel-Mouroz, 1972).

However, since the beginning of the 1960s, the forms of occupation and development of the coastal lowlands have developed considerably. This is because of the increased intervention of the State and public bodies in the processes of colonization and the creation of new population centres, leading to the introduction of political, social, and economic organization of the colonists, but principally because agricultural colonization is no longer the only pioneering phenomenon. Petroleum extraction, large dams, mines, and the industrial and tourist centres of the 'Frontier Resource Region' have widened the economic base of Mexico, attracting new populations and creating a new pioneer environment in which the town precedes rural development.

1. AGRICULTURAL COLONIZATION UP TO THE 1960s

It is 'a usable country of $180\,000$ km^2—the territorial equivalent of a second-rank European nation by surface area—which has brought historical Mexico to life' (Enjalbert, 1963). Apart from the indigenous centres of north Yucatan, highland Chiapas, and central Oaxaca, the principal population area and the principal cereal area was limited, at the end of the nineteenth century, to the *Altiplano* and the basins of central Mexico.

The opening-up of the arid coastal region and the humid tropics of the south by the railway (1880–1910), Porfirio Díaz's colonization laws, and the policy of large-scale irrigation followed by the post-revolutionary Governments, have contributed to the growth of this 'useful Mexico', and the redistribution of the main cultivated and settled areas.

1.1 Colonization and agrarian reform

After the phase of land speculation and monopolization by the land companies following the 1883 Colonization Law, agricultural colonization, from 1917 to 1962, played an essential, but ambiguous, rôle in relation to agrarian reform.

The 1917 Constitution does not insist on the return to the nation of all vacant land previously occupied by the Land Companies. Reserves of land were therefore available for private colonization, particularly in the south-east (Chiapas, Campeche, and Yucatan), and on the Gulf Coast (Veracruz and Tabasco). Official colonization remained small until 1940 with 153 colonies occupied by only 14 000 families on 1 200 000 hectares being created by the Presidents of this period. Moreover, colonization policy was mainly concerned with large-scale irrigation in the arid northern states, especially in the north-west. Between 1925 and 1947, 1.9 million hectares of irrigated land was reclaimed from the steppes of the north and from dry farming land. It is this effort by the Mexican State, with respect to irrigation, which explains why the land redistribution under agrarian reform did not lead to a collapse of agricultural production and to a food crisis in the towns.

Under Presidents Camacho and Alemán, who reversed the populistic agrarian policy of the Cárdenas Presidency, colonization became the means leading to real policies of agrarian counter-reform. Private property menaced by the overpopulation on the *Altiplano* reconstituted itself in the virgin land of the tropics. 'Fortunately for the future of the Republic there are areas along our coast with a prodigious potential that await only the creative power of men of action', declared President Camacho in 1941. The 1946 Colonization Law permits private colonization on national land, but while spontaneous colonization by *ejidatarios* gave only the right to parcels of land smaller than 20 hectares, the private colonist can cultivate up to 300 hectares of tropical plantation crops or can create cattle ranches fluctuating between 500 and 2500 hectares.

In twenty-six years of colonization, 1946–62, about 1200 colonies composed of 76 000 private landholdings had in this way occupied 6.5 million hectares by legal land allocation (Franco Bencomo, 1965).

By contrast, *ejidal* colonization, made possible by the 1954 Agrarian Code, creator of 'New Centres of Ejidal Settlement' (NCPE), only occupied 3.5 million hectares in 1966, and had only created 446 colonies (NCPE) with a population of 71 565 *ejidatarios*. The official effort in the *ejidal* sector remained modest, limited within its legal framework, while migration, the clearing of the forest, and now production, remained the domain of the spontaneous colonist.

1.2 Colonization and demographic equilibration

If one considers the tropical states between 1940 and 1960, it can be seen that the migration movement from the highlands outside these states played only a marginal rôle in the growth of the population. From 1950 to 1960, only 68 000 migrants came from the *Altiplano*, i.e. 4.2 per cent of the total population growth of the seven Gulf coast and south-east states. Moreover, states such as Oaxaca, Chiapas, and Yucatan lost more migrants than they gained. The 'Advance to the Sea' is largely a myth, which is confirmed by the stability of the relative demographic position of the seven states in the national population: 22 per cent in 1940 and 20 per cent in 1960.

Agricultural colonization and migration towards the tropical lands barely compensates for the migration towards Mexico City and the remainder of the country. On the other hand, as a result of colonization, the humid tropics absorbed more than the national average of new agricultural workers, even though they had only a fifth of the national population. The seven tropical states took 26 per cent of the new agricultural workers in the period 1940–50 and 33 per cent from 1950 to 1960.

In the same way, the number of agricultural units increased more rapidly than in the rest of Mexico (see Table 6.1). On the whole, while there is not general demographic growth in the Mexican humid tropics, there is a change in the distribution of the agricultural labour force.

1.3 Colonization and the growth of agricultural production

The Gulf and south-eastern states, which contained 20 per cent of the Mexican population in 1960, provided 25 per cent of Mexican agricultural production

Table 6.1 Growth in the number of farm units in the period 1930–70

	Mexico			Gulf–Yucatan–Chiapas		
	1930	1960	1970	1930	1960	1970
Total	1 390 903 100%	2 870 238 100%	3 215 796 100%	207 867 14.9%	527 158 18.4%	644 852 20.0%
Ejido members	536 983 100%	1 523 796 100%	2 218 472 100%	82 773 15.4%	314 594 20.6%	483 723 21.8%
Private farmers	854 020 100%	1 346 442 100%	997 324 100%	125 094 14.6%	212 564 15.8%	161 129 16.2%

and, in spite of the relationship of the indigenous population to low productivity, and the constraints of a difficult natural environment, a state such as Veracruz was able to raise itself into second position for maize production and first for livestock farming. From 1950 to 1962, the area farmed in the humid tropics increased more rapidly than anywhere else in Mexico, with the exception of the north Pacific coast, another area of colonization (see Table 6.2). Whilst the agricultural areas of the centre and north increased their agricultural production by modernization and increasing their yields, the humid tropics continued to increase their land clearing.

The development of new agricultural land has been accelerated in recent years (1950–73), but has not been accompanied by sufficient progress in intensive agricultural techniques; the humid tropics continue to improve their relative position in agricultural area, but are held back by their level of production in comparison to the centre-west and north-west (see Table 6.3). One of the duties of

Table 6.2 Annual growth of the agricultural sector by areas in the period 1940–62 (percentages)

	North Pacific	South Pacific	Gulf–Yucatan	North	Centre	Mexico
Production of 37 products	9.2	5.2	5.2	4.8	3.5	5.4
Area harvested	6.1	5.2	3.5	1.6	0.6	2.3
Yield	3.1	1.2	1.7	3.2	2.9	3.1

Source: Hertford (1967).

Table 6.3 Relative importance of the major areas of agricultural land growth (percentages)

	Area harvested		Value of production harvested	
	1950	1973	1950	1973
Gulf Coast + South-east States[a]	15.0	23.7	26.6	19.2
Centre-west + north-west states[b]	16.9	22.7	16.8	32.0
Mexico	100	100	100	100

Sources: Censo Agrícola Ejidal, 1950; Información Agropecuaria SAG/DGEA, 1973.
[a] Tamaulipas, Veracruz, Tabasco, Campeche, Yucatan.
[b] Aguascalientes, Jalisco, Colima, Nayarit, Sinaloa, Sonora, Baja California Sur.

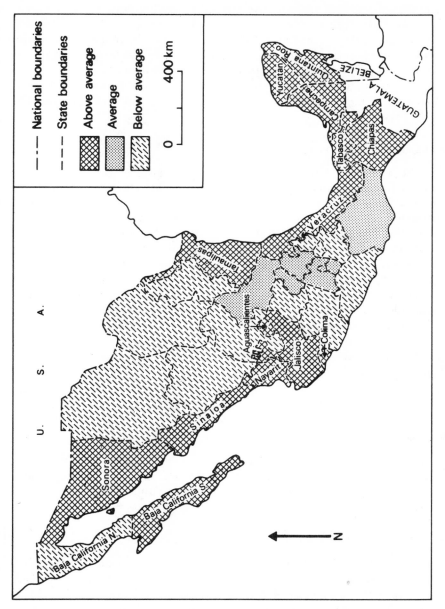

Figure 6.1 Mexico: growth of the cultivated area, 1950–73

official colonization in the more recent period is to substitute an intensive agriculture for the rudimentary extensive agriculture of the spontaneous colonists.

The states whose cultivated area has increased faster than the Mexican average are spread along the Atlantic and Pacific coasts from Tamaulipas to Quintana Roo and from Colima to Sonora, respectively, (see Figure 6.1). Out of the 6.7 million hectares of land that was newly cultivated between 1950 and 1973, 861 000 hectares are in Veracruz and 926 000 hectares are in Sinaloa–Sonora, which confirms the polarization of the colonization movement towards both coastal areas.

1.4 Colonization and large-scale development

Using the Tennessee Valley Authority (TVA) as an example, the Mexican Governments have created River Basin Commission, attached to the Ministry of Hydraulic Resources (SRH), with the object of dam, drainage and road construction improvements, which would permit regional settlement and development. The Papaloapan (1941), Tepalcatepec, Balsas (1947), and Grijalva (1951) Commissions provide the administrative framework for the 'Advance to the Sea' without, however, taking control of colonization, with the exception of the limited resettlement of displaced peasants. Thus, in the south of the Papaloapan basin, one of Mexico's important tropical frontiers, 42 per cent of the new colonization villages have been formed by resettlement and 55 per cent by the spontaneous colonization of the *ejidatarios* or small private owners, attracted by the new infrastructure. On the other hand, the role of the commissions has grown since the 1960s with the widening of development and colonization plans.

2. THE NEW 'ADVANCE TO THE SEA'

2.1 The new agricultural colonization policy

The turning point in colonization policy occurred in 1962, when the abolition of the 1946 Law meant that colonization of the national lands and of large estates affected by agrarian reform could only be carried out within the framework of the new *ejidos*.

2.1.1 The integration of colonization into agrarian policies

The growth in demand for *ejido* land encouraged the agrarian authorities to look to colonization as a substitute for the radicalization of the Agrarian Law, which would have implied a lowering of the threshold beyond which expropriation was automatic. President Lopez Mateos declared that, 'In order to solve the problems resulting from the shortage of land by grants in some parts of the country, we have undertaken a major operation to send peasants to virgin regions where land grants are possible' (Lopez Mateos, 1963). From 1965 onwards, the fall in the annual growth rate of agricultural production (from 5 per cent to 1 per cent) threatened the supply of food to the population and Mexico had to import maize. It also affected the capacity of the nation to finance its imports by agricultural exports. As the policy of intensifying agriculture and organizing the producers would take a long time, agricultural progress in the short and medium terms was, according to the economic planners, to be achieved by colonization. The aim was to bring under cultivation 375 000 hectares a year from 1970 to 1980, of which 100 000 hectares would be irrigated with small-scale irrigation schemes on the central plateau and large-scale irrigation in the north-west, 175 000 hectares would be cleared by public colonization (Programa Nacional de Desmonte) largely in the tropical zone to the south-east of the country, and 100 000 hectares of fallow and unused land was to be cultivated by private owners.

2.1.2 New methods of public colonization

Where the River Basin Commission and the Ministry of Hydraulic Resources took control of operations, directed colonization, which had multiplied under President Lopez Mateos, now became planned colonization, as in the Chac and Chontalpa Plans in 1962. Planned and regionalized colonization increased under President Díaz Ordaz and were institutionalized by President Echeverría. Colonization had become a means of developing national territory: 'to colonize is to rationally occupy all our territory' (Echeverría, 1970).

The new 1971 Federal Reform Law states that the elaboration and execution of regional plans for the creation of new population centres is in the public interest, and obliges Government departments to accept responsibility for *ejidal* colonization.

A Sub-Secretariat of New Settlement Centres, a branch of the new Ministry of Agrarian Reform, was responsible for carrying out the National Plan of Ejidal Colonization which was also put forward in 1971, with an Interministerial Commission for Ejidal Colonization to coordinate the work of different Ministries. The five fundamental objectives of this plan were (i) to redistribute the population towards areas having a production potential, but with few inhabitants; (ii) to create new sources of employment and income for the rural labour force; (iii) to bring about the rational and integral utilization of underutilized natural resources (or those threatened by serious misuse); (iv) to diminish rural–urban migration; and (v) to increase agricultural production by the cultivation of new lands (SRA, Sub-Secretaría de NCPE, 1976).

Because colonization is associated either with irrigation in the arid zone or with drainage in the low-lying tropical areas, the Colonization Plan is linked to the Plan Nacional Hidraulico which aims to acquire some five million hectares of drained or irrigated land from now until the turn of the century, in order to quadruple the 1972 agricultural production. These aims are derived as follows:

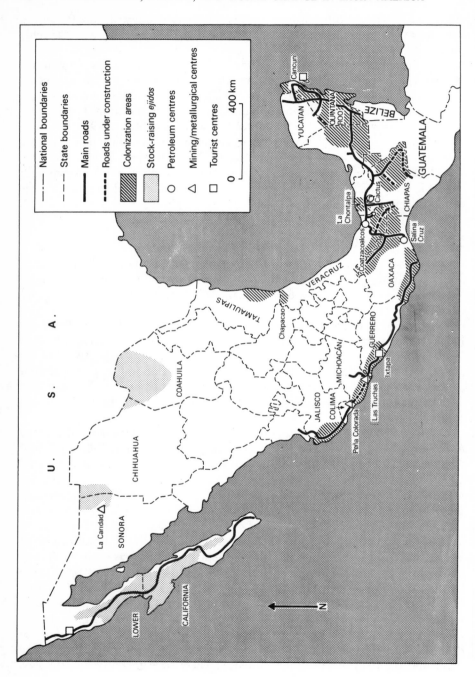

Figure 6.2 Mexico: active pioneer zones

	(million hectares)
Gulf–south east	3.3
Pacific–centre-north	1.5
North	0.2
Centre	0.5
Total	5.5

The large production unit is to be the instrument of colonization, by which *ejido* collectives would be created, so that the problems of agricultural modernization, insoluble with the present patterns of individual *ejido* holdings, would be resolved. The State aims for 'the reorganization of *ejidal* land, so that new methods of farming can be applied, allowing on one hand the consolidation of parcels to form a large-scale holding; and on the other hand, the introduction of new systems of work to facilitate the transfer of part of the surplus, produced by the modernization towards the other sectors where large-scale capital is concentrated' (Barkin, 1977). The collective colonization *ejido*, supervised by various agrarian, financial, technical, and administrative bodies (Ministry of Agrarian Reform, Banco Rural, and the Basin Commission and agencies such as that of the Chontalpa), should transform the colonization area into growth poles, creators of an exportable agricultural surplus, which would become the base for growth in local employment and consumption. These collective *ejidos* will be regrouped into larger operations to avoid the dissipation of the public effort.

2.1.3 Results

The National Plan for Ejido Colonization planned the creation of 112 NCPE on two million hectares, and 50 000 families each received land, although there are still 60 000 requests for land not satisfied, demonstrating the existence of two million potential *ejidatarios*. During the six years of President Echeverría (1971–76), thirty-seven projects were started distributing 1 485 000 hectares to 5051 families, and opening up 84 000 hectares for cultivation and 392 000 hectares for livestock farming. (Figure 6.2 shows the active pioneer zones.) The investment involved was in the order of five hundred million pesos without counting the annual credits for equipment and maintenance. The thirty-seven colonization centres produced goods worth 310 million pesos in 1975 (SRA, Sub-Secretaría de NCPE, 1976). Whilst part of this took place in Baja California and in the ranching *ejidos* of the Coahuila frontier area, the main area of official colonization was concentrated in the Gulf and south-east areas; namely, the 230 000 hectares Edzna (Campeche) project, the 250 000 hectares Ucum (Quintana Roo) project, at Xcan, at Cobaay in the Yucatan Peninsula, and particularly in the large-scale projects in the Grijalva–Usumacinta basin (see Table 6.4).

Since 1976, the Lopez Portillo Government has revived the policy of large-scale projects, in particular irrigation projects. 'The humid tropical zones consti-

Table 6.4 Agricultural–hydraulic projects in the Grijalva–Usumacinta basin

Project	Type	Surface area (ha)	Investment (million pesos)	State of advancement
Chontalpa I	Drainage	85 000	1 360	Completed
Balancán–Tenosique	Drainage	50 000	180	Completed
Cuatepoques	Irrigation	11 600	190	Outlined
Piamonte–Tacotalpa	Drainage	60 000	840	Outlined
Litoral–Tabasco	Drainage	45 000	450	Outlined
San Gregorio	Irrigation	10 000	153	Feasibility study
Chontalpa II	Drainage	185 000	2 960	Pre-feasibility study
Mezcalapa	Drainage	85 000	340	Identified
Los Ríos	Drainage	20 000	300	Identified
Sabana del Rosario	Drainage	60 000	200	Identified
Piamonte–Palenque	Drainage	120 000	720	Identified
Total		731 000	7 690	

Source: Azuaras Salas (1976).

tute virtually the only important agricultural frontier area which can grow in the future.... There are 3.6 million hectares capable of being cultivated. In a 50 000 hectare area we are already developing some six projects which represent development potential' (Lopez Portillo, 1978).

2.2 The different forms of colonization in the south-east

2.2.1 An integral development project—the Chontalpa (Tabasco) Plan

After the completion of several drainage and development projects, largely financed by the Inter-American Development Bank on the old Río Mezcalapa delta (1959–66), which made possible the construction of the Malpaso dam on the Grijalva, the Río Grijalva Commission decided to launch the Chontalpa Plan as an integral development project, rather than the smaller 52 000 hectares pilot project originally envisaged. The Chontalpa Plan contains a first stage (1966–76) with 83 000 hectares and 72 000 hectares to be developed in two phases (1966–1970 and 1971–76, respectively), followed by a second stage, bringing into use another 130 000 hectares. This plan involves both restructuring existing agriculture, which would affect about 8000 landholders, as well as colonization, with the clearing of the forest and the installation of 3098 colonists and their families.

The plans would have changed the agrarian structure, with 3000–4000 hectares units regrouping from small 10 hectares *ejido* parcels, and would also have totally modified the land, with the development of a drainage and irrigation canal grid. A new agricultural system would have been introduced, with the projected extension of commercial polyculture, (maize, rice, sorghum, and soya) oriented towards the interior market, and perennial crops for export, such as coffee, sugar-cane, rubber, and especially bananas, whose plantations would serve as the economic motor for development. The native population and the colonists would be grouped in twenty-two new villages with all social services provided, a village being situated in the middle of each agricultural unit.

However, from the outset, the application of the Plan met with opposition from part of the population, and the choice of agricultural products, as well as the Authorities' methods of intervention, had to be modified several times. Because of the lack of a real market, the banana plantations were limited, and, because of a lack of guaranteed annual finance, other cash crops were not introduced. The Plan was therefore reorientated towards cattle raising, which increased the size of the *ejido* plot and therefore reduced the number of beneficiaries, so that it was no longer necessary to appeal for colonists.

From the beginning, the change in the agrarian structure aroused the violent opposition of a group of local bosses, and the Army had to intervene in 1967. In addition, the creation of collective *ejidos* supported by the CNC (Confederación Nacional Campesina) union was resisted by *ejidatarios* who wanted to maintain the individual holdings and who found support from the rival CCI (Central Campesina Independiente) union.

These economic and social difficulties did not facilitate good relations between the different authorities involved in the Plan, which resulted in a delay in carrying it out, although in fact, the actual investment exceeded the foreseen financial limits. The aim of the first phase had not yet been achieved by 1975, six years behind schedule, with 1500 million pesos having been spent on it, almost double the 775 million pesos estimated for the 1966–70 period. Integral development was found to be expensive ($30 000 per family) and with poor results. The Mexican Government therefore decided to revise the Plan and, in 1970, it was decided partially to convert to sugar production, with 15 000 hectares to be planted serving a sugar mill with a capacity of 100 000 tons in order to step up production.

In 1972, the Chontalpa Plan Agency was formed to administer the Plan autonomously in association with the Grijalva Commission, which retained responsibility for technical and hydraulic tasks. The control of the *ejido* collectives was complete. Henceforward, the directors of the Agency established the planning of products, and controlled the credit and the technical assistance to the collective association, thus turning the *ejidatarios* into salaried farmworkers, in particular in the cane area that is dependent on the sugar mill. A temporary migrant labour force, recruited for the sugar harvest, every season enlarges the population of the new villages, a paradoxical situation because the *ejidatarios* are largely underemployed.

Table 6.5 Planned and actual agricultural development for the Chontalpa Plan, Phase I

	Before project 1964–66	Initial plan	Revised plan	Actual situation	
				1971–72	1974–75
Farmers	4 680	6 242	5 138	3 782	4 599
Private	2 090	2 090			
Ejidatarios	2 590	4 152	5 138	3 782	4 599
Area cultivated (ha)	35 922	94 000		33 785	49 188
Annual crops (ha)	8 640	48 000	12 250	3 884	14 907
Perennial crops (ha)	11 862	30 000	19 500	5 468	4 175
Sown pasture (ha)	15 600	16 000		24 433	30 106

Sources: SRH, Comisión del Grijalva, *Plan de la Chontalpa*; SARH, 1977 *Informe Parcial del Drenaje Agrícola de la Chontalpa*, Memorandum Técnico No. 368.

After a decade of uncertainty, the authorities are hesitating before starting the second phase of the Plan. Two major problems seem difficult to solve in the short term. First, it is far from clear what the newly cleared lands will produce, since the production plans for the first phase were not complied with. Secondly, there is the difficulty of deciding the method of production, since both *ejidatarios* and individual spontaneous colonists are suspicious of collective organization, blaming past failures on collectivization. The past experience of integrated development has led the outside agencies to take charge of all *ejido* activity. In all the colonization projects besides the Chontalpa, the process of outside bureaucratic control of agricultural production is continuing. The development of the plan is shown in Table 6.5.

2.2.2 Colonization and readjustment—the Uxpanapa Agricultural Plan

Between 1956 and 1957, following the construction of the Presa Miguel Alemán, the Papaloapan Commission moved the Mazatec communities found in the area of the reservoir onto the lands of old estates in Oaxaca State's Río Obispo colonization zone. However, due to the lack of land really suitable for agriculture, and above all as a result of a lack of financial support, the local colonists of this recolonization project could only survive by combining the collection of *barbasco* with the produce from 'slash and burn' farming. The relative setback of the project is demonstrated by the emigration of a group of colonists in search of new and better land (Ballesteros et al., 1970).

Following the large hydraulic investment of the Echeverría Presidency, the number of readjustment project experiments increased. The Río Grijalva Project was a source of tension among the local peasants affected. The *ejidatarios* affected by the Angostura dam (Figure 6.3) were to have been moved to new

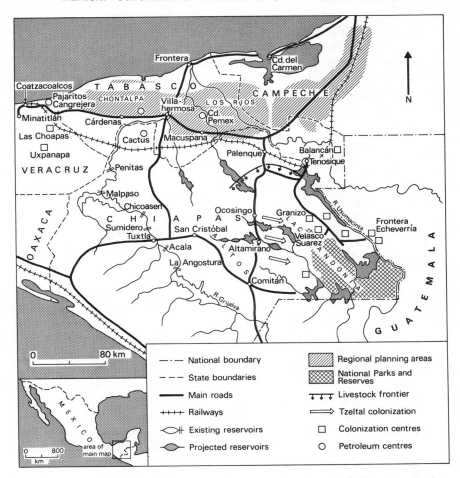

Figure 6.3 Colonization and development in the Grijalva and Usumacinta basins

ground to be cleared in the jungle but, refusing to leave their valley, they illegally installed themselves on livestock farmland on the surrounding slopes. The Army had to intervene in the Venustiano Carranza area in order to avoid conflict, and the Government sought to find a solution to the problem by buying back the land from the ranchers to give to the *ejidatarios*.

In these two examples, one realizes the hiatus between the official plans and the unforeseen reactions of the population, who were rarely asked their views about the future that the Authorities had planned for them. Since 1971, readjustment projects have represented almost a quarter of the colonization operation. The agricultural plans at Uxpanapa are the most important. The large-scale flooding of Papaloapan in 1969, provoked by the acceleration of man-made erosion in the eastern Sierra Madre and by the silting of its lower course, led President Echeverría to restart the improvement project of the Papaloapan basin, and, in particular, the construction of the Cerro de Oro dam. The fifty-one Chinantec *ejidos* affected by the project from 1973 had to choose out of three solutions: to stay in the neighbourhood of the dam, as did fifteen *ejidos*; to move to the Los Naranjos region, as did eight *ejidos*; or, as twenty-eight *ejidos* did, to integrate themselves into the Uxpanapa plan, to the south of Las Choapas in the Tehuantepec Peninsula (Allen 1979).

The Papaloapan Commission chose the Uxpanapa Forest area in order to reactivate *ejido* colonization, still in the embryo stage. One-hundred centres had been theoretically created between 1966 and 1974, covering 80 000 hectares, but only 1000 people migrated. The indigenous Chinantec colonists displaced by the Cerro de Oro dam became the new engine of colonization, with 4500 families expected. By Presidential Decree, the Distrito de Drenaje de Uxpanapa (covering 260 000 hectares) was created in January 1973 under the responsibility of the Papaloapan Commission who undertook to open up the area by constructing a penetration road from the main highway through the isthmus.

So, 12 000 hectares, of which 85 000 hectares were to be cleared for agricultural use, were reserved for indigenous colonists with fourteen villages with an average population of 1000–1500 each, containing 200 families planned. Each family head was to receive 20 hectares and the raw material to construct a crude house. From May 1974, the Papaloapan Commission assured the clearance of land, with an agency selling the cut wood to the railways, and the first colonists arrived. By the end of 1976, 10 000 hectares had been cleared and cultivated (maize, rice, pepper, and rubber), the *ejidal* colonists working collectively using machinery and credit from the Banco de Credito Rural. Nevertheless, problems arose due to the optimistic assessment of the potential of the land. Only 45 000 hectares were usable instead of the 85 000 hectares forecast, there were poor relations between the existing *ejidatarios* and the new indigenous colonists, and, finally, the change from an individual and manual to a mechanical and collectivized agriculture made the colonists totally dependent on the administrative and financial authorities. The medium-term success of the project depends on the continuation of work on the infrastructure, in particular communications, with 1000 million pesos to be invested, and on the consistency of the Authorities in sustaining the commercial cultivation of rice, pepper, or rubber, the economic bases of colonization (Secretaría de Recursos Hidraulicos, 1976).

2.2.3 *Ecology and the colonization policy—the Lacandon experience*

At the start of the 1970s, the Mexican scientific community started to criticize the harmful effect of uncoordinated colonization on the natural environment. The biologist, A. Gomez Pompa, denounced in the Press the Government policy of colonization by the destruction of the forests of the south-east: 'We avoid the scientific precedents and make bureaucratic improvization to solve today's problems, leaving the future generations with the problems of survival' (Gomez Pompa, 1970). The anthropologist, Fernando Benitez, showed how the large-scale development of Papaloapan had taken place without considering its effect on the indigenous economy of the hillsides based on the shifting cultivation of the Mazatecs and Mixes. 'The poor people of the basin were not considered, whilst the ranches of the politicians and the Generals flourished' (Benitez, 1970). But it is the current destruction of the Lacandon Forest which provoked the most passionate declarations. 'See the woods on both sides, the roads are burning . . . wherever the road goes the peasant follows, where the peasant goes, fire follows, where the fire goes the stones appear and where the stones appear stalks hunger' (Duby-Blom, 1970).

The rôle of the new roads, for example the Tuxtla–Pichucalco road (Figure 6.3), is condemned since, where the road penetrates, the sawmills overexploit the forest and the spontaneous colonists advance. The phenomenon is not particular to Mexico, and can be found in all tropical America. 'Without exception, government agencies lament the indiscriminate destruction of tropical forests and the irresponsible mining of the soil. For decades, however, they have remained impotent, and they may even have aggravated the situation with their road building policies' (Nelson, 1973). However, it is the spontaneous and nomadic colonization of the indigenous peasants (Tzetzals or Chols) coming from the Altos de Chiapas which is seen as the major danger. 'In Chiapas 2.5 million hectares of forest have gone, even so, the forest is a resource more important than petroleum', declared the Chiapas State Government, accusing those who invaded the forest of 'coverting the mahoganies and thousand-year-old cedars to cinders' (Velasco Suarez, 1976).

The Lacandon Forest, bounded to the west by the Altos, to the east and south by the Usumacinta, and to the north by the pioneer ranching zone in the Palenque area, is the land of the Lacandon indians, who presently number about 400 people. Up to the 1940s, the area was penetrated seasonally by the *monteros* and *chicleros* who collected the production of precious wood and chewing gum by the small towns of Tenosique and Ocosingo. However, from the start of the 1950s, a double pioneer movement became apparent; *ladino* colonists bought vacant land for extensive ranching—this being facilitated from Palenque by the construction of new roads—while to the south the Tzeltal indians, instead of taking part as before in seasonal migration towards the coffee *fincas*, permanently abondoned their villages (especially in the Oxchuc area) and descended the high valley tributaries of the Usumacinta, cultivating their *milpa* by the 'slash and burn' systems. This clearing of the forest by fire, sometimes on slopes reaching 50°, started an irreversible erosion process. The spontaneous indigenous colonizations, accelerated in the 1970s, led to the establishment of about 200 communities (70 000 people) in the Lacandon Forests, in addition to the 300 private ranches on the periphery (Port-Levet, 1979).

Faced with the threatened destruction of the forest within ten years if the colonization movement continued at the same rate and with the threat to the long-term potential of the Usumacinta hydroelectric power scheme, the Federal and Chiapas State Government intervened in 1974. The motives behind this State intervention appeared in a study undertaken by the Commissión de Estudios del Territorio Nacional (CETENAL). 'The large natural and cultural resources of this area should be used rationally through an integrated development plan with the aim of creating a development pole in this area which would surely influence the national economy in a positive way' (CETENAL, 1974).

The rationality of public intervention first of all showed itself in the legalization of the land occupation. Out of the 200 spontaneous colonist communities, about 120 received a definite grant of an *ejido* plot as new centres of *ejido* population, within the framework of the programme conducted by the Ministry for Agrarian Reform. A Presidential Degree in 1971 recognized the rights of the Lacandon indians who would have been 'in continuous, peaceful possession of these lands since time immemorial', to 660 000 hectares of forest. This measure took away the rights of spontaneous colonists already established in this area to practice shifting cultivation. The area thus legally protected, has been partially transformed into a Biosphere Reserve of 300 000 hectares created under the Unesco Man and Biosphere Project. This reserve, with the ecosystem placed under scientific control, has been extended to the Marqués de Comilla National Park, open to tourist use by private and public projects (Beltran and Halffter, 1976).

In 1975–76, the Chiapas State Government introduced a programme of displacement and concentration of the indigenous spontaneous colonists. Thus, 5000 Tzetzals were regrouped in 1978 in the centres of Velasco Suarez (Palestina) and 4000 Chols at Corozal, with the large villages on the Bonampak–Palenque road receiving the remainder of the colonists. This massive population movement was organized by the Authorities, who transported the colonists, gave them 25 pesos each per day during the installation period, and provisionally allowed them to harvest the *milpas* already sown.

While the survival of the natives, concentrated in the two villages situated in the middle of the protected area, depended on the introduction of new activities, such as aquaculture, fishing, tourism, and arboriculture, in the north another programme PIDER (Public Investment Program for Rural Development) sought to orientate the Tzeltal *ejido* towards growing coffee and rearing livestock. The promotion of coffee cultivation was made at the beginning of the 1970s by the Instituto Nacional Indigenista, who organized work from its centre located at Ocosingo. Because of the concentration of their life zone, the abandonment of 'slash and burn' for permanent agriculture, and the relocation of the cultivated land, the indigenous colonists have lost all autonomy in a movement directed by politicians, economists, and ecologists.

2.2.4 Colonization and the settlement of frontier areas

The creation of the NCPE 'Frontera Echeverría', on the Usumacinta, and the projected construction of a road in Lacandonia uniting Bonampak to Comitán, shows that strategic preoccupations underlie the control of colonization in Lacandonia. The concentration of natives in the support villages (such as Tres Naciones) bordering the Guatamalan frontier, recalls the situation observed in Venezuelan Guayana, where the National Frontier Commission settled the Pemon indians close to the Guayana Esequibo claimed by Venezuela.

The objective of territorial control and of national security in the areas where oil discoveries are transforming, or are going to transform, the jungle into a new El Dorado, can be attained by colonization, and economic profitability is only a secondary objective. In the Tabasco frontier area, and in Quintana Roo, large directed colonization operations were launched during Echeverría's Presidency. Taking a framework of colonization villages established during the Cárdenas period along the Río Hondo, which forms the frontier with Belize, the Ministry of Agrarian Reform has created large collective *ejidos* around the centre of Alvaro Obregon. Formed by 250–500 *ejido* colonists recruited by the Authorities in the northern and central states, and in particular the Laguna area, these villages have an economy entirely controlled and directed by the public bank (Banco de Credito Peninsular). The bank has imposed a highly mechanized agricultural system on the farmers, and wage has transformed the *ejidatarios* into labourers by the system of monthly 'advances' on the future earnings from the land. Although the recent creation of a State sugar refinery in this area can transform it into a regional agricultural development pole, the immediate aim is settlement of the Territory of Quintana Roo which has become a Federal State, and of its frontier area. This explains why loans for colonization are not solely concerned with work and production, although the *ejidatario* beneficiaries of this policy do not always understand the logic of it.

> Who knows how the Government works. When I was at Laguna, I had work from dawn to dusk and there was barely enough to eat. Now in these new villages we cross our arms and bank notes fall out of the sky.
> (*Ejidatario* quoted in Fort, 1979)

It is not just by chance that the organization of a large-scale project to develop the Los Rios region in Tabasco and Campeche is on the frontier: the Grijalva Commission has been administering since 1972 the Balancán–Tenosique Plan between the Río San Pedro and the Guatamalan frontier. The strategic objective is clearly defined by the Commission: 'The project assumes the establishment of our sovereignty in a jungle area with few inhabitants, on the frontier of our national territory with that of Guatemala' (Azuaras Salas, 1976).

The plan includes the development of 115 000 hectares (of which 65 000 hectares are to be developed between 1970 and 1976) and the installation of 3000 *ejidal* families from Veracruz, Michoacán, and Guanajuato. Each *ejido* receives 5000 hectares and a village is constructed for four *ejidos*. From the start, and for the first time in the colonization programme, a place was made for ecological research into the rational use of the forest and into selective reforestation: 'The necessary studies will be made to maintain as much as possible of the balance of the regional ecosystem, thus indicating a new way to plan agricultural developments' (Azuaras Salas, 1976). However, the clearing has, in fact, been extensive and only 300 hectares per *ejido* (6 per cent) of surface area has been conserved as forest, although agriculture (maize and rice) has remained marginal and it is mainly the grazing land which has been extended.

3. CONCLUSION: TOWARDS A NEW TYPE OF ACTIVE PIONEERING ZONE

Directed colonization and the control of spontaneous colonization movement is becoming more and more important in an overall and long-term policy of incorporating the humid-tropics into the national economy. The novelty of the 1970s is the appearance of a new generation of non-agricultural pioneering sites with centres of mineral extraction, petroleum complexes, and industrial and tourist poles overthrowing the geographical division of economic activities and of urban centres. The systematic development of primary raw material for the interior market and for export, as well as the policy of industrialization, have created the enclaves of Peña Colorada (iron-ore mines, Colima) and Las Truchas (steel-making at the junction of the Río Balsas on the Pacific coast).

But it is above all, the energy sector that has provoked the formation of multiple centres in the south-east. The hydroelectric development of the Río Grijalva has created large-scale construction sites attracting manpower for the Chicoasen and Angostura dam, and the petroleum boom which followed the discoveries of new reserves in Tabasco, Chiapas, and Campeche has led, since 1973, to the opening-up by roads of the south-east and a swarm of extraction and treatment units. The new Cactus and Cangrejera petrochemical areas, the construction of a giant refinery at Salina Cruz, and the multiplication of jobs linked to petroleum at Villahermosa, Cárdenas, Coatzacoalcos, Minatitlán, and Las Choapas has led to new currents of migration competitive with pioneer agricultural migration. In the same way, the policies of building large seaside tourist complexes to attract foreign visitors, by FONATUR (Fondo Nacional del Turismo), has given birth to the new centres of Cancun (Quintana Roo) and Iztapa–Zihuatenejo (Guerrero).

The new urban centres (250 000 people at Coatzacoalcos, Minatitlán, 70 000 at Lázaro Cárdenas–Las Truchas, 30 000 at Cancun) offer new markets for food products. The economic profitability of integral colonization projects lies

Table 6.6 Demographic growth and migration in the areas of the humid tropics in the period 1960–70

Zones of the Comisión Nacional de los Salarios Minimos	Population		Growth 1960–70		Estimated net migration 1960–70
	1960	1970			
Centre and south					
Pacific coast	1 493 718	2 165 959	670 241	44.9%	101 134
Colima–Manzanillo	77 874	134 227	56 353	72.4%	26 683
Guerrero–Acapulco	84 720	238 733	153 993	181.8%	121 715
Gulf and Caribbean					
coast	3 820 444	5 429 298	1 608 854	42.1%	153 265
Veracruz–Minatitlán	122 739	243 675	120 936	98.5%	74 172
Campeche–Carmen	47 183	84 211	37 028	78.5%	19 051
Quintana Roo	50 169	88 150	37 981	75.7%	18 867

perhaps in an orientation of production towards the new urban markets, and not solely with the search for exterior markets.

The 'Advance to the Sea', from now on, will increasingly become an 'Advance to the Town' and towards employment in the secondary and tertiary sectors. Moreover, the migration of the 1960–70 period had already assured something of the character because amongst the pioneer zones (those whose population has increased markedly) one finds Acapulco (tourism), Minatitlán, and Poza Rica (both petroleum)—see Table 6.6.

This industrial and tourist growth has re-established a certain demographic and economic equilibrium between the humid tropics of the Gulf and the Pacific, up to now handicapped by the lack of large areas of virgin land to clear.

REFERENCES

Allen, E., 1979, 'Infrastructural investment and new settlement. The Papaloapan basin, Mexico', in *L'Encadrement des Paysanneries dans les zones de Colonisation*, Travaux et Mémoires de l'IHEAL, No. 32, Paris.

Avila Camacho, 1941, *Presidential Speech*, July.

Azuaras Salas, E., 1976, 'La comisión del río Grijalva, su contribución al desarrollo regional'; in *Desarrollo de la Cuenca Grijalva–Usumacinta*, Mesa Redonda de la Secretaría de Recursos Hidraulicos (Mexico: Instituto Mexicano de Recursos Naturales Renovables).

Ballesteros, J., Edel, M., and Nelson, M., 1970, *La Colonización del Papaloapan* Mexico: CIDA).

Barkin, D., 1977, 'Desarrollo regional y reorganización campesina. La Chontalpa como reflejo del gran problema agropecuario Mexicano', *Comercio Exterior*, **27**, 12.

Beltran, F., and Halffter, G., 1976, *Programa Interdisciplinario Para el Estudio Integral de la Selva Lacandona*, CONACYT.

Benitez, F., 1970, *Conferencia al ICMI*.
Blom, G. Duby., 1970 'Chiapas Testimonio y documentos', *El Dia*, 28 March.
CETENAL, 1974, *Estudio de Gran Visión de la Zona Lancandona*, Mexico.
Echeverria L., 1970, *Informe de Gobierno*, 1 December.
Enjalbert, H., 1963, 'Le milieu naturel et le Méxique utile', *Tiers Monde*, **4**, 335–59.
Franco Bencomo, J., 1965, *Los Nuevos Centros de Población*, Serie Monografías No. 4, Escuela de Agricultura, Centro de Economía Agrícola, Mexico.
Fort, O., 1979, 'Structure agraire et encadrement des paysans dans les secteurs de colonisation de l'Etat de Quintana Roo'; in *L'Encadrement des Paysanneries dans les Zones de Colonisation*, Travaux et Mémoires de l'IHEAL, No. 32, Paris.
Gomez Pompa, A., 1970, Articles in *Excelsior*, 22 July.
Hertford, R., 1967, 'Una evaluación de mis primeras impresiones', *Seminario Sobre la participación del Sector Agrícola en el Desarrollo Económico de Mexico*, Mexico.
Lopez Mateos, 1963, *Informe de Gobierno*, 1 September.
Lopez Portillo, 1978, *Informe de Gobierno*, 1 September.
Nelson, M., 1973, *The Development of Tropical Lands. Policy Issues in Latin America* (Baltimore: Johns Hopkins Press).
Port-Levet, F., 1979, 'La selva lacandona Chiapas, Méxique: controle de l'espace et des productions et encadrement de la paysannerie indigène'; in *L'encadrement des paysanneries dans les zones de colonisation*, Travaux et Mémoires de l'IHEAL, No. 32, Paris.
Revel-Mouroz, J., 1972, *Amenagement et Colonisation du Tropique Humide Méxicain. Le Versant du Golfe et des Caraibes*, Travaux et Mémoires de l'IHEAL, No. 27, Paris.
SRA, Sub-Secretaría de NCPE, 1976, Article in *El Día*, 12 April.
Secretaría de Recursos Hidraulicos, 1976, *Memoria de la Construcción de la Obra en Uxanapa*, Comisión del Papaloapan, Mexico.
Velasco Suarez, M., 1976, *Speech*, 1 October.

Environment, Society, and Rural Change in Latin America
Edited by D. A. Preston
© 1980 John Wiley & Sons Ltd.

CHAPTER 7

Rural innovation and the organization of space in southern Brazil

RAYMOND PEBAYLE

In comparison with the north-east which is, with some difficulty, breaking free from its constricting heritages and from a pioneering and predatory Amazonia, the Brazilian south* is now recovering from a virtually complete conquest of its territory. However, while one might imagine that the south is stable and more rustic than dynamically pioneering, closer examination shows that the area is still influenced by the *bandeirante* spirit and is experiencing real change. These changes permeate not just rural life but the organization of space in general.

Our intention here is to determine the exact relationships existing between rural innovations and regionalization at a time when new technological inventions are combined with traditional pioneer-style migration to create, perhaps only temporarily, new spatial structures. With this in mind, after having shown the limiting factors of agricultural innovation, we shall try to demonstrate some of the characteristic types of changes and their impact on the regional organization of subtropical Brazil, a map of which is shown in Figure 7.1.

1. THE FACTORS OF CHANGE IN THE RURAL SOCIETIES OF THE SOUTH OF BRAZIL

In the customary environment of the farmer and stockbreeder of the south of Brazil, there is a group of more or less limiting factors for the adoption and spread of innovations. Some of these factors are certainly peculiar to a subtropical area. However, it would seem that this region has been isolated far too much from its national context, to the point of wanting to make it an exception, whereas it shares with the rest of the territory and people of Brazil a number of common characteristics.

Broadly speaking, one can distinguish three sets of conditions for innovating action in the country areas of the south.

---

*By south or subtropical Brazil is meant the States of Rio Grande do Sul, Paraná, Santa Catarina, and southern Mato Grosso (the Dourados region).

Figure 7.1 Southern Brazil

(i) In the first place, there are the *antecedent conditions*, inherited and therefore difficult to change. The land situation is certainly less restricting here than in the north-east, nevertheless, apart from the fact that landless people also exist in subtropical Brazil, one cannot fail to be impressed by the frequency of smallholdings, today to be found concentrated almost exclusively in forest areas. This is partly the heritage of voluntary colonization of lands originally abandoned by the stockbreeders of the grasslands. In these areas, the large estate (latifundium), dating from the early *sesmarias* distributed by the Portuguese Crown (1 league by 3 leagues in size), has been rapidly decreasing in importance

Table 7.1 Land ownership in Brazil and the southern states (percentages)

		Farm size category (ha)					
		< 10	10–100	100–200	200–500	500–1000	> 1000
Brazil	Farms	51.1	39.2	4.4	3.1	1.0	0.8
	Farmland	3.1	20.5	10.1	15.7	11.3	39.1
São Paulo	Farms	45.9	43.8	4.9	3.5	1.1	0.8
	Farmland	3.3	25.1	12.0	18.1	13.2	28.2
Paraná	Farms	53.2	43.5	1.8	1.1	0.3	0.2
	Framland	10.7	41.6	9.3	12.1	7.4	18.9
Santa Catarina	Farms	31.6	64.1	2.4	1.3	0.4	0.2
	Farmland	4.7	51.0	9.5	11.4	8.3	15.0
Rio Grande do Sul	Farms	34.6	58.8	2.8	2.1	0.8	0.6
	Farmland	3.6	32.4	8.4	14.2	12.6	28.9

for some decades. In contrast, Table 7.1 shows the relative importance of middle-sized holdings, considered locally as average (10–100 hectares). This is undoubtedly a fairly typical characteristic of the south.

The land situation, however, is the most confusing factor of evolution of a peasant society. Although he is an innovator here, the large landowner can be considered, in other places, to be an unbearable reactionary. On the other hand, caught up in an ultrarestrictive land network in the innermost settlements, the smallholders of European origin, who grow many different crops, can easily make use of very small holdings if they live near a large market which encourages them to cultivate profitable crops.

It is still the case, however, that the large landowner is always in a better position than the small farmer to carry out a major innovation. For, apart from the fact that people only lend to the wealthy, in Brazil as elsewhere, the large landlord can make use of the resources of indirect exploitation to introduce fundamental changes. In these cases, innovation entails practically no supplementary expenses for the landlord, who merely accords rights of possession to sharecropper who puts the innovation into effect and shares its eventual advantages.

On the other hand, there is tradition. Paradoxical in a new country, tradition is established very quickly in an isolated rural society. Among the heritages, there is one, in particular, that has for a long time bordered upon the mythical: the very elementary determinism which closely associated forest and agriculture on the one hand, and grassland and stockbreeding on the other. Thus, as long as trees and grass have been so rigorously separated in their agricultural use, no fundamental change has appeared in the south of Brazil.

A further tradition is the pioneer mentality, which has expressed itself by migrations in the country between the original settlements of the south and the vast empty forest areas of the centre and the west of the States of Santa Catarina and Paraná. These movements, which are now spreading over the Mato Grosso and even Amazonia, correspond to a certain conception of the economic viability of land as much as to acquired habits. We shall return to this later.

Finally, we have the socio-cultural heritage. The south is certainly the most European region of Brazil, because of the ethnic origins of its inhabitants. But, apart from the fact that it would be foolish to try to attribute any superiority to the Central European element over the Portuguese descendant, it certainly seems that acculturation is very advanced in the colonies of the south of Brazil, despite the stubborn survival of dialects which bear only a remote resemblance to their languages of origin. The fact remains, nevertheless, that cultural isolationism (a formidable obstacle to change) has often made some rural societies, which have remained isolated from the major centres, opposed to progress.

(ii) At the *local level*, change is firstly conditioned by the agricultural qualities of the soils, and also by the natural dynamics of the rural communities. On the first point, it must be emphasized that the southern territory of Brazil is certainly the most favoured part of the nation for one essential resource: water. For it shares, together with the forest (*mata*) of the north-east and the mountains of the Atlantic, this invaluable precondition for innovation by tropical and subtropical peasantry. Indeed, when it is present for at least six months of the year and is regular from year to year, water facilitates the introduction of changes so that one can dispense with the very costly irrigation arrangements which are so greatly lacking in the north-east, for example. It also permits a considerable choice of cultivated plants and stockbreeding in an environment to which man can easily adapt himself. In this respect, one cannot fail to be struck by the extraordinary correspondence that exists between the areas of Brazil well provided with water (with the exception of Amazonia) and the highest incomes per capita.

But, do change and innovation find, in the midst of the rural groups of the south of Brazil, the motivation essential for their adoption and diffusion? *A priori*, this rural society still seems ill equipped to make good use of fundamental innovations. Although the south is by far the most literate region of Brazil, complicated innovations are assimilated with difficulty by the isolated individual. This latter, perfectly aware of the risks involved in the isolated adoption of an innovation, prefers to follow the methods based on a new plant or form of stockbreeding which can seem highly profitable. But even when a new technique is easily assimilated, one quickly becomes aware of the first alarming signs of monocultures—the exhaustion of natural environments and the fluctuations of prices depending on the double hazard of internal speculations and the international economic situation.

But the last few decades have witnessed considerable modifications in the initial emergence of innovations. Indeed, the community leaders are changing. Traditionally, they had in their favour fortune, connections, the experience of a great age, or the spiritual investiture of a religion. Today, the *coroneis*, like the *padres*, are tending to lose their influence. The present period is one of the transferring of prestige, slow but undeniably unfavourable to the former leaders. Thus, in stockbreeding societies, the most respected figures are no longer the traditional political bosses, but the landowners who have travelled and therefore learned a great deal.

Another novelty is that the State is now establishing its own technical and financial counsellors. It also grants loans, with land as security. Having for a long time been suspicious of engineers and technicians from outside the community, the farmers and stockbreeders of the south are beginning to follow their advice, especially if it is accompanied by promises of substantial agricultural loans. Thus, for at least a period of ten years, those officials in charge have been associating controlled credit with the adoption of the innovations regarded as indispensable in the rural areas.

(iii) Finally, there are the *subsequent conditions*. They comprise the markets of agricultural products and, naturally, access to them. Thus, if undeniable progress has been made in the last ten years in the field of road transport, it is nevertheless true that vast areas of south Brazil still remain economically dominated by a few large companies, national or foreign, which take advantage of the isolation of the producers to purchase, at a very low price, crops which they will sell at an extremely high price on the national and foreign markets. The wholesalers are often also speculators. Sudden 'shortages' of essential foods (oil, haricot beans, rice, meat, etc.) on the large urban markets are everyday occurrences in Brazil. The State itself can become a speculator, as was the case in 1975 for coffee. One can appreciate that such speculative dealings hardly encourage innovation, and a lot of small owners, disappointed by some unsuccessful attempt at commercial production, today prefer to devote themselves to mixed food farming.

Certainly, there is the domestic market which one knows to be constantly growing. Urban growth, in particular, has stimulated the demand for some essential agricultural products, but precisely because they are in part foodstuffs, these products do not create very elastic demands. In any case, the national market is certainly incapable of absorbing the surplus commercial crops which remain closely associated with foreign rather than domestic demand.

Finally, one can see how rural innovation cannot, under any circumstances, be divorced from the entire social and economic context, and therefore from a certain regional organization of territory.

In two out of three instances, changes are still very much influenced by *pioneirismo*, that is the pioneer mentality by virtue of which the farmer, convinced that there are new lands towards the north or the west, largely unoccupied and cheap, rejects the idea of the permanency of the peasantry which he regards as uneconomical. For a great number of rural areas in the south of

Brazil, many changes are regarded precisely in this pioneer perspective. One can appreciate, moreover, that, in a country where almost two-thirds of the area is still empty or only slightly occupied, the first form of change is not technological or local, but consists rather of a simple spatial transfer of a form of farming, which is well tried but predatory and therefore temporary.

This is the most classic form of pioneer reflex which exists. Nevertheless, in the case of the south, things are both more complicated and more logical than a straightforward nomadism, which would tend to assimilate the colonists and the planters to become common itinerant farmers. For the movement towards the pioneer fringes is not, in fact, a flight of men who are abandoning a totally exhausted natural environment. Indeed, the movement is always combined with certain forms of innovations in the lands of emigration, which are only very rarely deserted.

According to the reasons for departure and the changes of the old land, one can just distinguish two pioneer movements: one consists of men who come to look primarily for a refuge in the new lands; the other consists of movement for reasons chiefly of economy and advancement, of men looking for an El Dorado in an unoccupied new land.

2. COLONIAL PIONEER DYNAMICS AND REGIONAL SCLEROSIS

The uninterrupted advance of the colonists of European origin from the extreme south of Brazil towards the forested lands of the centre-west of Santa Catarina and Paraná corresponds fairly closely to the first type of migration between rural areas.

In fact, from the middle of the nineteenth century right up to the present day, occupation of the subtropical virgin lands of the three states of the south (with the exception of northern Paraná) has taken place in two ways. The one, official, consisted of the settlement of foreign colonists whose initial establishment was facilitated, at least in principle, by the State. Nuclei of settlers were thus formed on the escarpment and the southern border of the large basaltic plateau (Rio Grande do Sul), in the valleys of the coastal rivers of the east of Santa Catarina, and in various parts of the Araucaria forest of the central plateau of Paraná. In their turn, these initial settlements have brought about spontaneous pioneer migrations. In fact, if there has actually been an itinerant agriculture in the Rio Grande do Sul, it did not involve the colonists of German, Italian, or Polish origin. It was carried out by authentic *caboclos* who were the first inhabitants of the forests, with the indians, and whom colonization, as much official as private, has totally ignored until the early years of this century. Constantly driven back towards the north by the colonial pioneer fronts, these occupants, who were denied all legal existence, were almost itinerant farmers. In particular, they practised a very elementary system of fattening of pigs, following the so-called *safristas* method. This consisted of the casual cultivation of a maize field, which was given over to the pigs when harvest time was approaching. The presence of these *caboclos* close to the pioneer fronts explains the frequency of *capoeiras*, or bush-covered fallow grounds, that interrupt the forest cover, which is in theory virgin and intended for the real colonists.

The behaviour of the colonists is far removed from the nomadism of the mestizo populations that preceded them. Indeed, for the colonists of the south, the forest areas of plateaux (Araucarias) and valleys (rain forest) have always been natural areas for expansion. The systems of farming and occupation of land of those people practising mixed farming in the colonies are adapted to this availability of fresh territory. In fact, originally possessing lots varying from 80 ha (in the last half of the nineteenth century) to 25 ha (in the beginning of the twentieth century), the very prolific colonial families, and those using systems of cultivation requiring the land to be left fallow for long periods, have been short of land after one or two generations, according to the period and the soils. They then had the choice of three solutions: exodus, the adoption of more intensive systems of cultivation, and in-migration towards new lands. Of these three solutions, the most logical and the most immediate always seemed to be the third, as exodus was limited by the shortage of jobs in towns, and the local adoption of new systems of cultivation would have meant financial sacrifices and risks which people could not or were not prepared to undertake.

Thus, innovation in these societies of isolated mixed farmers has not generally been of a technological nature, but has consisted more of spatial transfers of traditional systems of exploitation. People moved to new areas, in the knowledge that the new lands of the north were both more fertile and of better value than those in the south.

From this, it is easy to see how innovaton, in its proper sense, has been constantly rejected by this rural society which was able to recover its land capital, qualitatively and quantitatively improved, solely by means of migration from one forest to another. The entire system of occupation of the forest lands of the Brazilian south, to the south of the 24th parallel, has been based on these long waves of migration between country areas, motivated basically by the elementary demographic increase of colonial families.

Such a system of dynamics presented, nevertheless, two dangers. One lay in the ageing and the impoverishment of the population of the pioneer's original country. For the departures towards the new lands have been especially by young couples whose far-sighted parents bought for them, whenever possible, new forest lands in the north. In these circumstances, the colonial holdings of the south were left in the hands of older peasants, lacking both capital and the desire to innovate. The second danger has been obvious for less than a decade: the lack of new and cheap lands is beginning to be felt. Now, it is necessary to go further north, and already several thousand colonists from the south have tried their luck in Amazonia, sometimes well before the recent opening of the trans-Amazonia roads. The colonists of the Paraná are for their part trying, so far with success, to settle in a foreign territory, in eastern Paraguay.

Today, therefore, innovation is imperative in the south of Brazil. But how can one innovate when one is a penniless smallholder, little inclined to follow the advice of agricultural experts and economists whom one mistrusts as much out of ethnic tradition as peasant suspicion? The changes adopted are, in fact, fairly straightforward, too simple even to have any notable influence on the regional life of the south as a whole. According to a typology which we have already put forward (Pebayle, 1974), we shall say that these colonists have only rearranged, sometimes renovated, their traditional systems of agriculture without ever crossing the threshold of a real agricultural change. Thus, it is a simple reorganization of the existing elements of exploitation that causes the chief innovation of most of the colonies of the south of Brazil: the rearing of pigs to be sold to the local slaughterhouses for cold storage, especially to those of São Paulo. In this case, the colonists of the south have merely acquired more pigs and fed them with the surplus cereals, roots, or tubers which they had previously sold as best they could at the local markets or to unscrupulous intermediaries. Soya cultivation was adopted because it was so straightforward, since this leguminous plant was introduced in the traditional *roças* without appreciably altering the customary system of grouping together food crops. A new row of soya was simply added to the lines of maize, haricot beans, manioc, or rice.

Here, then, is an elementary principle of pioneer dynamics which has brought about a total absence of true agricultural changes. From this, it is easy to appreciate that the timid colonial country areas have not generated a very enduring regional life. This is already seen by the small number of regional centres that they contain (Figure 7.1). The colonial region of the south of Brazil is indeed served by a few centres whose commercial functions would not be capable of justifying real economic polarization if recent banking networks and some regional offices for rural development had not reinforced incomplete commercial functions. For, of the nine regional centres picked out by the Brazilian Institute of Geography and Statistics (Fundacão IBGe, 1972) in the forest areas which are today almost entirely occupied by colonists cultivating many different crops, only two (Blumenau and Caxias do Sul) have been classified as regional centres of the first order. The seven others (Ijuí, Santa Rosa, and Erechim for Rio Grande do Sul; Chapecó, Joaçaba, and Joinville for Santa Catarina; and Pato Branco for Paraná), are proving totally incapable of ensuring the maintenance of trade and services in spheres of influence which are often more illusory than real. Moreover, 30 to 40 per cent of all exchanges effected by these regional centres with their environment are made with centres of an equal or inferior standard in other regions.

In fact, São Paulo, especially, Pôrto Alegre, and Curitiba—this latter to a lesser degree—control the destiny of these still badly structured areas. The first two metropolises, in particular, are also the two markets that absorb the bulk of the pork and cereal products of the south of Brazil. They also allocate credits, initiate fundamental innovations, and monopolize decisions. The industrial superiority of São Paulo, and the actual monopoly it exercises over the administrative authorities, is such that no change can be accomplished in the subtropical area of Brazil without Paulista backing. It is significant that the only relative exceptions to the ascendancy of this large national metropolis are two colonial towns which have been able to find, in their traditions as much as in their ethnic cohesion, the strength necessary to create an original secondary sector and certain independent commercial services. Blumenau, in the old German colonies of Santa Catarina, is thus a well established regional centre, thanks to its industries, especially textiles. Caxias do Sul, an Italian town, has profited both from the stabilizing rôle of a long-lasting and commercial crop—vineyards—and from cooperative initiatives capable of standing up against the total control by a few large firms dealing in viticulture and the sale of wine.

It is precisely the permanent and commercial cultivation of coffee in south-east Brazil which has brought about the other form of pioneer dynamics, sometimes less paralysing when it comes to innovation and regionalization.

3. AGRICULTURAL INNOVATIONS AND THE PIONEER SUCCESSION OF THE PLANTERS

The coffee-growing pioneer front is not, like the food-producing fringes of the extreme south of Brazil, a simple expansion in space of a unique system of cultivation. It appears, on the contrary, like the spearhead of a chain of transformations which, within a few decades, modify the natural environment and create a totally new ecological situation, which sometimes brings about the establishment of actual regions. In these conditions, man is certainly at first a predatory pioneer, but he can rapidly become an organizer of space.

Held back by the conquest of new lands in the agriculturally diversified south, innovation accompanies pioneer expansion, when this has the powerful motivation of a very profitable plantation. For the new lands, ecologically suited to coffee growing—in appearance, at least—have given rise to a conquest of the west where the pioneer has looked more for an El Dorado than a refuge. And, behind the battlefront formed by men in search of wealth, the post-pioneer area remains inert for only a short while. Indeed, the wealth produced by the plantation, and the local accumulation of capital, very quickly gave rise to an intense urban life which featured a number of banking and commercial services linked to the plantation and the commercialization of coffee. It is only a short step from this stage to the rapid establishment of total control of the area. Certain towns have been able to succeed but not without occasional catastrophes.

The recent development of the north-west of the Paraná is a good example of the processes of change peculiar to the coffee-growing pioneer fronts. Although intended for small planters and originally made up, in fact, of *sitiantes*, the pioneers' march from the new north of Paraná brings to mind that of the *fazendeiros* of São Paulo. As in this latter State, the coffee-growing front features, in the early stages, rapid sequences of clearing, plantation, adaptation, and urbanization. For the purely predatory phase of the pioneer front (which cuts down and burns the forest, in order to sow, in apparent disorder, coffee, grains, and some food crops) is succeeded by rotation of crops more suited to the soil. Very quickly the plantations, the food crops, and the pasture lands spread right down the slopes. Simultaneously, indispensable tertiary activity is established. In the Paraná, the sites of the towns were largely established in advance: in São Paulo, on the other hand, the urban centres rose up spontaneously. Whichever the case, the processes of urbanization are the same as those described by Monbeig (1952): the banker, the wholesaler, and the coffee-broker set themselves up in charge of urban development which is made easier by the local formation of capital.

Thus, the departure from classic pioneer predation quickly reaches the stage of agricultural change and gives rise to intense urban activity. From the early years, the pioneer coffee fringe thus originates innovations which effect all activities to do with settled territory.

It is precisely at this stage that one often encounters two catastrophes which are practically part of the processes of change and regional growth in the hinterlands of the coffee-growing pioneer fronts. These catastrophes consist, first, of the crisis of overproduction, and the brutal revelation of an ecological inability to adapt, in the form of a frost or an early exhaustion of the soil. These two catastrophes have affected the new north of Paraná in cycles, whereas São Paulo has only known crises of overproduction.

Each of these cases results in a new organization of space, a new link of what we have decided to call the pioneer succession. Throughout this phase, change does not only affect the agricultural environment; according to the physical morphology, and equally on the forms of land tenure, innovation transforms the planters' world to a greater or lesser degree. Of the planters, the *sitiantes* are more ready to choose departure towards new lands or a return to cultivating known crops, rather than costly radical changes. The big *fazendeiro* planter, on the other hand, will opt for total change if he considers the catastrophe to be too serious. In this way, many sandy regions of São Paulo and the Paraná have known a sudden return to stockbreeding as soon as the sandy soil—and sometimes frosts—have convinced people of the coffee plant's inability to adapt ecologically on the western borders of the two States. It is true that this change has been made easier by the great number of wage labourers, sharecroppers, and colonists left without work by the crises. These men accepted temporary and unprofitable grazing contracts, by means of which the owners succeeded in replacing the coffee plants by tropical fodder (*Panicum maximum*). This conversion cost the future stockbreeder practically nothing, since, in payment for his labour, the sharecropper only received the right to occupy the land for two or three years, during which he was to share his produce with the owner.

Finally, between the stages of ultraconservatism and sharp change, comes an agricultural conversion which replaces plantation by annual crops. Thus, northern Paraná and part of the Paulista west have for some years adopted a large-scale mechanized cultivation of wheat and soya. This is found on soils with a basalt base, where loans from the Bank of Brazil have made it possible to envisage the eradication of old or frozen coffee plants, without risking a disastrous decline in production.

Present-day landscapes are a striking example of these post-pioneer changes, not only of the new spatial distribution of agricultural activities, but also the changes, often brutal, to towns and their domination of the rural environment. Indeed, whereas the urban centres of the post-pioneer districts, populated by smallholders incapable of breaking with their recent past devoted to planting, are stagnating or slowly becoming deserted, the towns of the areas of rich *terra roxa* are undergoing complete change and becoming increasingly active. New technologies, new produce (soya in yielding excellent harvests), and new people too, are bringing about spectacular commercial recoveries (agricultural machinery, piecework and repair centres, etc.), as well as banking and industrial recoveries (oil refineries). Unfortunately, innovation has also had its social consequences by releasing the masses from the coffee industry and throwing them, unemployed and starving, out of the towns. These men will henceforth constitute the workforce of day labourers employed on a daily basis and deprived of the guarantees offered to full-time employees.

But the day labourers are even more numerous in the regions changing over from planting to stockbreeding. And, in this case, their situation is worse still, for their summary eviction from their lands has not had the immediate result of new jobs in the city produced by the other changes. In fact, in the new stockbreeding regions, the towns are dying out, deprived both of the industrial and commercial services brought about by coffee and of the everyday services justified in the past by the presence of the abundant workforce of the plantation. Today, in the sandy soil of the sandstone area of Caiua, for example, scarcely one stockman is needed where thirty labourers were not enough on the coffee plantations.

The present polarization of the area of São Paulo and northern Paraná is a good example of these unequal transformations of post-pioneer territory. Thus, immediately to the west of the Paulista peripheral depression, one sees a regional network clearly taking shape in areas where post-pioneer agricultural changes have been many and varied. From Ribeirão Prêto and São Jośe do Rio Prêto, in the north, as far as Londrina and Maringá, in the south, the towns of Araraquara, Bauru, Ourinhos, and Marília comprise, together with the preceding centres, a well developed urban network. On the other hand, where agricultural change has been tentative or submerged by the enormous stockbreeding *fazendas*, regional life is still very slightly organized, as in the two regional centres of Araçatuba and Presidente Prudente, which serve somewhat loosely large areas, which have very little interaction with regional urban centres. Here regionalization has been quite simply paralysed by rural change which has not stimulated human development nor a more dynamic pattern of land ownership.

One can thus see how pioneer development, apparently uniform and innocuous, can influence in very different ways the future of vast tracts of land according to the socio-economic origins of its instigators and the strength of the agricultural innovations. These changes determine an entire series of post-pioneer transformations, thanks to which, in the most favourable instances, distinct regions are created within a few decades. But such transformations are still exceptional, for the pioneer still has so little impact on the spatial organization of land which he has exhausted and that has been abandoned in favour of new areas, or on which he develops extensive systems of stockbreeding, from which the day labourer is excluded.

4. URBAN-ORIENTED CHANGE AND THE DEVELOPMENT OF THE GRASSLANDS

The pioneer dynamic is certainly not totally absent from the changes that we are going to describe. They are even obvious in certain forms of expansion by crop farmers determined to innovate and conquer. Nevertheless, these changes are no longer only related to an elusive El Dorado which exhausts the environment and fails to accommodate men for long. On the contrary, this time it is a question of a new style of settlement, in which inter-ethnic contacts and the actions of the townspeople or bankers create the changes and in this way bring about new forms of regional life.

Three recent types of evolution provide a good example of the emergence of the new southern Brazil. The market-gardening and dairy industries, and the small stock farms have naturally experienced some changes in the vicinity of the large towns. However, one would be wrong to imagine them as always being very specialized forms of farming which ignore the traditional systems of working the land. In fact, it is often because of a reorganization of farming systems in the country areas that the urban centres receive their daily supplies. In one place, someone will rear hens; in another, more work will be devoted to the kitchen garden; further on, someone will try to fatten a small herd of cattle. Thus, some surplus goods are produced for despatch to the nearest town. In this respect, it must be stressed that these small suppliers of the urban markets are not necessarily installed near the town. In such a huge country as Brazil, a tarred road is more often than not sufficient to cause a special line of business to grow up many kilometres away from a market, whereas one would more readily imagine it to be situated on the outskirts of a town. On the other hand, when there is a bad road which impedes access to farms, food crops are grown largely on every edge to the town.

Nevertheless, truly specialized farming does exist. It is then less scattered and often carried out by very distinct ethnic communities. Thus, Curitiba's milk is largely supplied by colonies populated by Mennonites from Central Europe, one colony of which is Witmarsum, situated about 80 km from the capital of Paraná. Generally, however, the milk is supplied from farms that have partially or totally modernized their traditional methods of farming. Some specialities remain under the control of certain foreign communities; for example, the dessert grapes grown by the 'Italian' municipalities of Jundiai and São Roque, in the State of São Paulo. The Japanese market-gardeners and poulterers are even more noteworthy for, with an exceptional family workforce and just as remarkable business sense, they practically dominate the supply of vegetables, eggs, and poultry not only for the São Paulo market but also a good number of urban centres of the south. The sense of solidarity within the community can be seen in the very powerful cooperative formed by the Japanese of São Paulo to avoid the pitfalls of the traditional networks of distribution.

Oriented towards satisfying urban demands, these innovations have obviously not modified the pre-existing regional organization. Their impact on the countryside is limited to a very irregular circular zone on the outskirts of the towns and at isolated points along the major roads.

Although more dispersed, the innovations introduced in the distant country areas, at the instigation of several large agricultural processing industries have, in general, modified the traditional way of life very little. It is known how rapidly such industries can spread important cultural innovations throughout the colonial countryside of the south. The cultivation of barley thus depends on several large breweries, that of fruit in the region of Pelotas also depends on a few dozen industries installed in the town. But the most outstanding example is that of the Souza Cruz Tobacco Company, alias the British–American Tobacco Company. This enterprise has given rise to a special form of small peasant farming in several municipalities of the German colonies of the south. Dictatorial in its technical and financial framework, in complete control of the domestic market, the Souza Cruz Company presides over the destiny of huge areas which it has sometimes brought to the brink of destitution after having led them to believe in an agricultural El Dorado. It is thus not surprising that, in general, such innovations have hardly inspired a very vigorous urban way of life.

The second example of change, the irrigated cultivation of rice of Rio Grande do Sul, is to date a unique instance in Brazil of control of water and natural environment over several hundreds of thousands of hectares (350 000 hectares in 1970). However, this fundamental change in the open countryside of the south of Brazil has not entailed any remarkable changes in the functional organization of space. This is obvious when consulting Table 7.1, where one can see that, with the exception of Santa Maria and Pelotas, no regional centre of any importance exists in the flat lands of the south of Brazil—the valleys of the Rio Jacui and Rio Uruguay, and the shores of the lagoons of the coastal areas.

This curious anomaly is, in fact, associated with both human and physical conditions dominating the development of rice cultivation in the *varzeas* of Rio Grande do Sul. It was a meeting between the traditional landowning stockbreeders, on the one hand, and the townspeople or advanced farmers, on the other, which gave rise, at the beginning of the century, to the first attempts at irrigation by pumping. After the initial success, the association of stockbreeders and farmers made its existence official by means of an agricultural contract known as a *parceria*. According to this contract, the stockbreeder retained ownership of the *varzeas* where rice was cultivated, which he leased to the farmers for a certain time only. The farmers payed their rents in kind, by handing over part of their harvests. It is important to note that despite the profits gained from this new crop, and despite the financial, technical, and commercial aid given by the State to the rice growers, they did not succeed in becoming owners of the lands. In 1970, for example, more than two-thirds of them were still tenant farmers. The stockbreeders refused to sell land, even though it was not good for livestock, but it was capable of providing a very comfortable and reliable annual income, when rented to farmers specializing in irrigation.

In these conditions, the cultivation of rice in Rio Grande do Sul has brought about a situation which is completely paradoxical in the Brazilian context. On the one hand, it is the only activity to have resulted over large areas in any far-reaching control of water. But, on the other hand, it has remained a nomadic cultivation because of the maintenance of systems of rental which have limited the length of the farmer's stay on the stockbreeder's lands.

It is consequently easier to understand the functions of the towns in these rice-growing areas. In general, their population has noticeably increased, because the rice growers live there. Their commerce has acquired new activities (machinery, fertilizers, etc.), and they have become the centres of active cooperatives, and the offices of the Bank of Brazil are all-powerful. However, with few exceptions, these towns do not dominate their municipal areas, and only very occasionally do their spheres of influence extend beyond the immediate zone of the town. The explanation is simple: the shifting cultivation of rice causes the growers to settle not in the country but in the towns. Only a fairly poor and unstable proletariat and a few permanent employees of the stockbreeding *fazendas* live in the country areas which have practically no local urban centres. All the tertiary functions are thus deprived of room for action, since they are carried out exclusively between those who live in the town. In the case of the cultivation of rice in Rio Grande, agricultural change, which is essential from the technical point of view, has not resulted in very perceptible progress of the organization of space. The main factor responsible for this situation would seem to be the land ownership situation.

The essential rôle of the land tenure system in rural innovation and the regional organization is demonstrated by a third example of agricultural change, typical of subtropical Brazil. This is the instance of the large mechanized farm of the uplands which has practically transformed the economy and society of the grassland areas from Caçapava and São Borja, in the south, as far as Dourados, in the Mato Grosso.

The initial stages of crop farming in the pasture lands of the gauchos are well known. They were strongly encouraged by the Government; cultivating only wheat, from 1947 to 1956, the new farm of the plateau had very logically to endure the classic failures of any monoculture: continuous cultivation, with no resting period or rotation, which facilitated the increase of various devastating illnesses which practically ruined the new farmers towards the middle of the 1950s. Paradoxically, it is the stockbreeder who bore the brunt of this crisis. For here, too, the farmer was initially an intruder to whom the stockbreeder leased lands on payment of rent. However, for wheat growing, rent was payed in cash and not in kind, as the early harvests were proving extremely hazardous. Thus, when the crisis occurred, and the necessity of remedying the almost desperate situation of the tenant farmers, one of the solutions adopted was to declare a moratorium on the payment of debts and rents which were frozen for several years. In these conditions, those who lost most were the landowning stockbreeders who received, after a great delay, rents that were ludicrously low because of the terrible inflation of the 1950s.

The real change was due to the financial discomfiture of the traditional landowners of the grasslands and the constantly renewed aid, given by the Bank of Brazil to the tenant farmers. Land passed increasingly into the hands of the arable farmers, who were economically strengthened by the adoption of a new crop, soya. These land transfers and the profits of agriculture explain two essential events of the human geography of the south of Brazil. The first consists of a veritable neo-pioneer change which, originating in Rio Grande do Sul, is at the moment transforming the grassy savannas on *terra roxa* to the south of Campo Grande. Throughout, one can always note the same loss of land by the stockbreeders and the establishment of the new method of farming. This is an extraordinary phenomenon if one remembers that everywhere else in Brazil the ascendancy of the stockbreeder is still unchallenged.

A second consequence is that agricultural change is certainly responsible for the beginning of the growth of an urban network, until now unknown in the stockbreeding lands. Well established in centres like Passo Fundo, Cruz Alta, and Santo Angelo on the plateau of Rio Grande, urban development is now crystallizing around Ponta Grossa on the plateau of Paraná. In the south of Mato Grosso, the town of Dourados literally mushroomed—in 1976; 305 commercial establishments enlivened a town whose population had more than doubled in 10 years to have some 54 000 inhabitants in 1975. The distributing rôle of Dourados in a countryside undergoing complete change is shown by certain of its sales: 30 tractors sold in 1971, 460 in 1975! The number of businesses connected with rural life was three in 1962; 59 in 1975! Simultaneously, an enormous wheat and soya cooperative included the new farmers of most of the neighbouring municipalities, while a refrigerated abattoir, under construction in 1977, promised to complete the town's receiving rôle over a huge area to the south of the State of Mato Grosso.

This last example is significant, for it enables an assessment of the precise consequences of technological innovation carried out in a favourable environment. Here, in effect, the mechanized farm succeeds the stockbreeding *fazenda*, which is huge but unprofitable. The new farm is financed by the Bank of Brazil, its owners are experienced, and its expansion is perfectly compatible with the pioneer mentality and the geographical environment.

Compatible—this word would definitely seem to be fundamental to a study of innovation. We already know that, at the level of a rural farm, an innovation is more easy to adopt if it is compatible with the agricultural system in use. It is the same for the introduction of the change in an inhabited area, with however one difference; unlike the individual who categorically rejects an innovation deemed to be incompatible with his farm, the region can remain, on the whole, fairly indifferent to the impact of a purely agricultural change. Its introduction, compatible only with the immediate interests of a well defined group or social class, can prove to be ruinous or crippling for all the other groups or classes of the inhabited area, and therefore for the survival of their regional organization. This is precisely the case with the large stockbreeding farm on *Panicum maximum* which one today finds as much in the south as in Amazonia.

From this, one perceives that only a systematic search for compatibility with what already exists would be able to harmonize innovation and inherited spatial organization. This is not very often the case. The methods of achieving this would, moreover, imply a series of changes or parallel adaptations which prove practically impossible to realize in Brazil's present economic, social, and political circumstances. It would be necessary, for example, to count upon more united communities (like, for example, certain Italian colonies of the south), whereas we know the traditional communities to be very loosely knit, as has been clearly shown by de Queiroz (1976). It would be necessary to be able to lessen the control of the large farms and of commercial monoculture intended for foreign markets. We are well aware that the present economic policy of Brazil is aimed at just this sort of exportable produce and the large-scale farming devoted to it. It would, finally, be necessary to control all the area available, whereas we know that it is all not yet opened up and the land regarded in a very pioneer fashion.

BIBLIOGRAPHY

Fundacão IBGE, 1972, *Divisão do Brasil em Regioes Funcionais Urbanas* (Rio de Janeiro: Instituto Brasileiro de Geografia).
Monbeig, P., 1952, *Pionniers et Planteurs de São Paulo* (Paris: Armand Colin).

Pebayle, R., 1974, 'Une typologie de l'innovation rurale au Brésil', *Les Cahiers d' Outre-Mer*, **108**, 338–355.
de Queiroz, M. I. P., 1976, *O Campesinato Brasileiro* (Petropolis: Livraria Vozes).
Rogers, E. M., 1958, 'Categorizing the adopters of agricultural practices', *Rural Sociology*, **23**, 345–354.

Cultural Change

Environment, Society, and Rural Change in Latin America
Edited by D. A. Preston
© 1980 John Wiley & Sons Ltd.

CHAPTER 8

Cultural change and ethnicity in rural Mexico

LOURDES ARIZPE

An unease with the term 'cultural change' seems to be characteristic of research in rural areas in Latin America in the last few years. It stems, in my view, from two sources: first, from the confusion arising between the holistic term of culture as used in classical anthropology, and the reductionist version of it prevalent in other social disciplines and in policy-oriented studies, and, secondly, from the belief that the concept of cultural change cannot be dislodged from the North American culturalistic framework, and thus cannot be applied in studies taking a Marxist or dependency theory framework.

Because these theoretical difficulties have not been made explicit, very different approaches have been taken in different studies. In some, cultural change has been subsumed into a more general process of 'modernization'; in others, it has narrowed in focus into ethnic change; and in others still, it has been left aside, since the implicit assumption is made that the ideological superstructure (that is, culture in a restricted sense) is nothing more than a mechanical reflection of the economic infrastructure, and thus efforts should be directed at explaining the latter.

This paper attempts to give a comprehensive view of cultural change in the Mazahua region in central Mexico, since the beginning of this century, by focusing on the association between mainstream economic processes and shifts in ethnicity, in literacy, and in the perception of rural culture in the region. In so doing, it is hoped that both narrowing down the concept and separating it from the economic and power structure will be accented.

Studies of cultural change in Latin America indeed began with a culturalistic approach. The Malinowskian scheme of culture change was a major influence and was applied in Mexico (Foster, 1960). Later, cultural change came to be known as 'acculturation', a term meant to bridge the gap between otherwise static ethnographic accounts of cultures. But 'acculturation' could not occur unless it was assumed that cultures with discrete boundaries existed which were becoming other cultures. Thus, indian cultures were assumed to exist, as direct inheritors of a pre-Hispanic past. Under the weight of such assumptions, however, it became increasingly clear that the social groups and processes that could explain the direction, intensity, and selectivity of social change were lost (for criticism of this approach see Stavenhagen, 1974; Pitt-Rivers, 1973; Van den Berghe, 1974; Ribeiro, 1968; Bonfil, 1972).

In the attempt to overcome the difficulties posed by this culturalistic approach, modernization theory has attempted to provide a comprehensive scheme of change from 'traditional' societies to 'modern' societies, which could be applied to all rural areas undergoing changes around the world (Lerner, 1958). This dual scheme for societies in the Third World, though, has been severely criticized (among others, by Singer, 1975; Ribeiro, 1975). The main difficulty, it seems to me, is that modernization theory does not add any analytical insights. It narrates a process and divides it up into diverse stages, again and again. But even if it is divided into a thousand stages, it will still be the same scheme, one which, bowing to Redfield's heritage, fundamentally assumes economic change. Redfield intended to provide a blueprint for *social* change with his folk–urban continuum, but instead he provided one for *cultural* change in a restricted sense of the term. It was not the *technical* order that concerned him: he assumed that the peasant economy would by and large evolve into an 'urban' type of economy. What interested him was to explain the *moral* order. In the same way, in modernization theory it is assumed that traditional peasant economies are being integrated into national and international market systems. Of course, there is this constant transformation going on, but the vital issue to explain is, surely, why changes occur *differentially* between nations, regions, and villages, and along class lines, ethnic lines, and sex lines. Thus, my proposition is that, once the major economic trends of capitalism in peripheral countries are known, attention should turn from trying to explain the uniformities of this process to explaining the disparities within it. This, it seems to me, is especially important in the relationship between rural and urban areas at the local level.

Unequal development at all levels, from the local to the international, is a consequence of the dynamics inherent in capitalism. At the local level in rural areas, this theoretical approach has tended to postulate a mechanical relationship between core and periphery which has already been contested, correctly in my view (Roberts, 1974; Kemper, 1970), but no alternative analysis of the relationship of culture to mainstream economic processes has arisen. This relationship is a highly complex one and cannot, at this time, be understood in overall terms. The first steps in this direction will have to be to point at areas of contradiction and of reciprocal influence.

That cultural change in rural Latin America has closely responded to events on the national scene is becoming more and more evident in historical studies. For example, in Mexico, already in the eighteenth century, the apparently isolated rural areas responded to the widespread economic recession by the formation of closed, corporate indian communities that consolidated a 'traditional' way of life (Wolf, 1959). Seen from this perspective, traditionalism, as several authors have noted, becomes a symptom, rather than a cause. However, at a theoretical level, it is necessary to postulate that at certain historical periods cultural phenomena can, in fact, overdetermine economic processes.

In the Mazahua region in the last fifty years, cultural change has gone ahead in stops and starts, differentially for ethnic groups and for social classes, with geographical disparities between villages. Although one same trend is evident at the regional level, this paper analyses the disparities and factors underlying it.

In this paper, the terms 'culture' and 'culture change' will be used throughout in a restricted sense, referring primarily to modes of thought, language, art, customs, and rules of social behaviour.

THE MAZAHUA REGION

The Mazahua region encompasses eleven municipalities in a high arid plateau 300 km north-west of Mexico City with a total population of 200 000 of whom roughly half speak the Indo-American Mazahua language. It is an agricultural region of *ejido* lands (lands owned by the State allotted to the peasants), with an acute fragmentation of land and low productivity that is leading to a rapid proletarianization of the peasants. As a result, both the mestizo and the Mazahua households are involved in seasonal and permanent out-migration, mainly to Mexico City and the Mexico–United States border. Migration and the spread of mass media and literacy have brought a strong cultural impact to the region, building up pressure for the changing of values and cultural habits. Cultural change in the last fifty years, though, has occurred differentially along ethnic and class lines.

ETHNIC GROUPS IN THE REGION

At the turn of the century, the social structure of the region, whose pivots were the export-oriented *haciendas*, clearly allocated rôles in production along ethnic lines. It may be even more correct to say, according to what informants stated, that ethnic identity was allocated by position in the production system. The owners of the *haciendas*, informants say, were all of Spanish descent. In fact, only one family, De la Fuente y Parres, owners of La Providencia, the largest *hacienda*, was of such descent; other *hacendado*'s families were assigned this putative descent because it was appropriate to their identity as the landowning class.

The foremen and administrators were mestizos or impoverished whites. Only one informant mentioned this distinction, all the others considered them as mestizos. Thus, no difference was made between these two groups, either in social or in cultural terms. The boundary drawn between them and the labourers and peasants, however, was strikingly clear—the latter were indians. That this distinction was in no way racial is shown by the fact that children, such as the illegitimate offspring of the *hacendado*'s sons, and the mestizo's legitimate or illegitimate offspring—frequently resulting from the rape of Mazahua women—who were brought up in the *hacienda* household became mestizos. Those, of the same parental origin who were raised among the Mazahua became indians.

The Mazahua, in turn, called and still call the other group mestizo or *cruzado*—'mixed'—and call themselves *mazahueros*. They refer to their nearest geographical and linguistic neighbours by the name of their language: the Otomi—to the east and north—and the Matlatzinca—to the south. They consider them very different from their own group, and only rarely will they call them *hermano indio*. The mestizo, in contrast, make no distinction among these groups, and call them all indians. This coincides with reports from other places such as Guatemala (Van der Berghe, 1974).

In *hacienda* times, there was no distinction between the *peones acasillados*, the Mazahua labourers who lived on the *hacienda* premises, and those who lived in peasant households in the villages. The former were a specialized workforce who laboured in the workshops where the root of the *zacatón*—a grass—was processed for export to Europe and the United Staes.

The Revolution of 1910 came to the region entirely from the outside. The *hacendados* fled abroad, the mestizos entered the fray, and the Mazahua fled to the nearby hills. By 1925, the only major apparent results of the fighting in the region had been deaths, famine, and epidemics. The *haciendas* had missed a few steps but still continued production. It was not until the Federal Government decided to distribute the land that the mestizos, organized into *agrarista* groups, invaded the *hacienda* lands.

By the end of the 1920s and in the early 1930s, the Government had allocated *ejido* lands to almost all the villages in the region. Significantly, having obtained the lands, the mestizo and the Mazahua populations agglutinated into new settlements along ethnic lines. At this point, it seems to me, ethnic loyalty overrode other considerations. If the creation of the ethnic identity of the indian was fostered and perpetuated by assigning Indo-American peoples a distinct economic rôle in the colonial economy, and later on in the *haciendas*, the fact that Mazahua cohesion persisted after these pressures subsided shows that their ethnic identity was by no means only an economic specialization. Interestingly, the data show that, from that point in time on, a dialectical relationship was established between ethnic filiation and economic pressures.

The Mazahua villages got their land and immediately retracted into corporate, isolationist units based on a household economy of the same type that Wolf (1959) describes in rural areas in eighteenth-century Mexico. It is significant that, in almost all Mazahua villages, this coincided with the rise of *caciques* who deliberately and ostentatiously proclaimed their Mazahua identity. In the mestizo villages, in contrast, the *caciques* doubled as local and regional representatives of the ruling party and new entrepreneurs in trade and in commercial agriculture. Thus, the economy of these villages rapidly underwent the changes attendant on the spread of capitalism into this rural region.

In the 1930s, both ethnic groups received equal amounts of land, and capital goods were distributed more or less equally between them. By the 1970s, the Mazahua were clearly the poorer of the two. Elsewhere (Arizpe, 1978), a detailed analysis is presented of the social, economic, political, and cultural mechanisms that help explain their progressive impoverishment. All peasant households in the region have felt the impact of economic forces that destroy the peasant economy, but for the Mazahua the consequences have been more acute. Women, especially, have lost sources of income in traditional handicrafts, cottage industries, petty trade, and traditional occupations such as midwife and *curandera* (healer). Men also lost occupations such as plough carver, musician, bone setter, and others, as well as wage labour in the mines and in the *hacienda* fields. Sources of cash income for the households declined while their needs for cash soared. The sons and daughters of mestizo families, because they benefited from schooling and from social priority in entering employment, have taken on the new occupations generated by capitalist development such as seamstress, tailor, teacher, and electrician, and have taken up new economic opportunities to set up small business such as shops, restaurants, and taxi and bus services. Since so many of them left agriculture, their brothers—women were legally barred from inheriting *ejido* lands until 1975—were able to inherit sizable amounts of land. In contrast, the Mazahua younger generations are locked into the agricultural sector and have had to inherit ever more fragmented plots of land. This has led the domestic units to more acute decapitalization and proletarianization.

At the national level, important cultural events exerted a strong influence in the region. At the end of the 1920s, José Vasconcelos crystallized a new concept of the Mexican nationality: the *raza cósmica*, a cosmic race that would successfully blend the Indo-American and the Spanish heritages. Since the indo-American groups were allocated *ejido* lands with no ethnic distinction being made, the new Government expected 'acculturation', a term just then beginning to be used in academic circles, to occur naturally and automatically. As the data from the Mazahua region show, in spite of this open invitation, the indian groups remained indian. In the 1940s, as it became increasingly apparent that these indo-American communities not only were not integrating into the economic dynamic of the new society, but continued to be as exploited and repressed, at the local level, as ever, the Instituto Nacional Indigenista was established. The avowed purpose of this institution was to integrate the communities into the national economic and social system. This policy of opening indo-American communities in order to achieve development by, among other things, enlarging the internal market, couched itself in cultural terms and, as will be seen further on, was responsible for building up pressure to make the indo-Americans reject their indian identity.

Although there was no direct action of the Instituto Indigenista in the Mazahua region, the policy of *indigenismo* (indianism) made itself strongly felt, as can be seen in this excerpt from an interview with a Mazahua man: asked what he thought about the fact that the Mazahua language and customs were disappearing he answered:

> But, lady, isn't that what the government wants? Because we speak here but only here with those who live in the village. We should rather speak as they do elsewhere, as the nation does. So we shall make a fatherland.

Further on in the interview, he stated explicit reasons for wishing his children to learn Spanish:

> This would allow them to do better; oh yes, he (his child) can go to ask for a job and the boss asks him—'what knowledge do you have?'—and if he has no knowledge they don't give him the job. On the other hand, if he can express himself, they say to him—'come here and get yourself some pennies.'

Ethnic filiation also influenced the strategies taken by the peasant households faced with progressive proletarianization. Migration became the most important source of income for both groups, with offspring being increasingly sent off, by order of birth, to wage labour. Mestizo young men migrated at first towards the Mexico–United States border to work as agricultural labourers in the United States. After the *bracero* programme (the official treaty between the two Governments) was terminated in 1964, they went to Mexico City where they were generally able to get jobs in factories, in offices, or in the services sector as bus or taxi drivers, agents, or salesmen. Mestizo young women also went to Mexico City and usually got jobs as shop assistants or in offices. Many of them, both women and men, settled permanently in Mexico City.

In contrast, during the 1950s, the Mazahua men began migrating in search of agricultural work to various regions and to Xochimilco, an agricultural suburb of Mexico City. They rarely settled in these places, since their primary aim in migrating was to get additional income for their parents' household. In the 1960s, they began to migrate increasingly to Mexico City to work as market porters, masons, and as assistants in La Merced market stalls and warehouses. The Mazahua women, almost without exception, worked as domestics and rarely had occupational mobility. They usually left the job to get married, usually to someone from the village.

Again, the fact that the Mazahua migrants normally returned to their village increased population pressure on the land within that group. Some Mazahua villages experienced permanent out-migration. In the 1950s, Mazahua migrants were able to to escalate into formal permanent employment because of the overall expansion of the manufacturing and services sector in Mexico City. In the 1960s, their mobility slowed down as the expansion waned. Skill and documentary requirements increased for factory work, and the Mazahua then began to stay within their ethnic community in the city. That is, the dialectic that made 'passing' into the urban culture advantageous came to a standstill.

In terms of ethnicity, two strategies emerged to cope with unemployment and occupational immobility. First, the community specialized in a trade activity. Mazahua women intensified their street selling and retained their Mazahua identity because it became useful in competing with other city vendors. Second, many of the migrants joined a new religious brotherhood, the *Concheros*, in what seems to me to be an effort at recreating ethnic solidarity—and through it, financial and social reciprocity—in an otherwise fluid and unstructured cultural context.

A NEW URBAN ETHNIC IDENTITY

The *Concheros* originally came together to dance purportedly Aztec dances at church celebrations in Mexico City. The group, named after the *conchas* or shells they use to accompany their music, became a brotherhood in which initiated members address each other as *hermano* and *hermana* and engage in highly ritualized and hierarchized social relationships.

With an ease of creativity that other social classes in Mexico society certainly deny themselves, these people have blended together Aztec symbols and designs, ritual elements of the *mayordomías* and *cargo* systems of some of their rural villages, Catholic religious liturgy, and the oratory and pomp of political ceremony, to create a new, eloquent image of their collective identity. Their costumes are sometimes luxurious and beautiful, their conviction and enthusiasm contagious. Their new identity is probably more genuine than that of the urban middle classes who derive theirs passively from the mass media.

Most of their members are of rural origin, with a great number of Otomi and Mazahua among them, and many of them have lived in the city for fifteen years or longer. The men work at low-income jobs such as street vendors, doormen, lottery tickets salesmen, nightwatchmen, and the like. The women do domestic labour as well as part-time informal activities, street selling, sewing, and washing clothes. For almost all of them, the city has not kept its promise of providing a permanent, well paid job. Caught between the rural cultures they left behind and the urban culture dominated by the middle class, and which is out of their reach, it seems to me that they have to revert to purported indian and Aztec symbols to form an immediate reference group and a new identity. In the older *barrios* or *vecindades* of migrants of Mexico City, such subcultures have always been created. For example, the subculture of the *peladito*, so well portrayed by Cantinflas, the Mexican comic actor, or the subculture of 'Tepito'. In the case of the *Concheros*, the fluid residential mobility and the new shanty towns have not been able to provide a *barrio* basis for a new subculture. Thus, they have formed their own by reactivating rural ties of before, and the Aztec culture.

Although urban-based, *Conchero* groups frequently go to dance in the villages of origin of their members. For example, they danced at the patron saint's day fiesta in Santiago Toxi, a Mazahua village.

In the view of the urban middle class, they belong to an indian, 'backward' culture. In fact, it is a *new* indian subculture created for specific purposes of cohesiveness and solidarity in a social space that has nothing to offer but a cultural vacuum.

Interestingly, many *Concheros* are literate and have had some schooling. In fact, it is from school textbooks that they have extracted elements of the Aztec culture to use as their new symbols.

SCHOOLS IN THE REGION

The rôle of the schools in cultural change in the region can be illustrated by data from two Mazahua villages. Primary schools in Santiago Toxi and San Francisco Dotejiare opened in 1945. In both, Mazahua children stayed away from the schools in Toxi until the late 1950s and in Dotejiare until the early 1960s. The reasons cited for this were that schoolteachers and mestizo children discriminated against, and often mistreated, Mazahua children. Also, informants said, the children were needed to help in the homestead. Boys and girls had to tend the flock of sheep, girls had to help with the domestic work. Female children, in particular, were not allowed to go to school, not only for the reason just cited but because it was feared that they would be raped, a fear totally based on real possiblities. Even today, young girls are taken out of school as soon as they begin to menstruate. And also, it was thought that women did not need an education.

Among the mestizo families, both male and female children attended school. The rape of mestizo girls is less frequent, because of the brutal retaliation that men in their families would exert. Mestizo parents were always very conscious of the advantages inherent in acquiring skills, and sometimes sacrificed their need for additional labour in the home so that their children could attend school.

These two attitudes could be interpreted as stemming from different cultural traditions. But the differential access of the two ethnic groups to employment and to economic opportunities show that both were rational decisions within the context of each group's sphere of action. Mazahua children could not enter wage employment in the services of offices in the villages or towns. School curricula were strikingly urban in content so that, in fact, they could not apply what they learned at school to their agricultural activities. The mestizos, in contrast, could apply this knowledge, since they would be dealing with national institutions and with jobs that required skills learned in the schools. Moreover, for Mazahua children, attending school usually meant a change in ethnic filiation, since teachers forbade them to speak their language and to wear Mazahua garments.

Since they were to live in the villages, such an estrangement from the Mazahua community was detrimental to their chances for survival and mobility.

This is supported by data from Toxi showing that Mazahua children began attending school massively during the second half of the 1950s. It was in this village that population pressure on the land was greatest, so that migration to Mexico City began earlier than in other Mazahua villages. Also, families had little use for all of the labour of children, having lost their herds and subdivided their lands. Success in wage labour in Mexico City, unlike the local situation, depended on a good knowledge of Spanish and basic skills in reading and arithmetic. For example, a young woman or man who could keep accounts (*hacer cuentas*), got a much higher wage than other helpers who only carried merchandise or waited on customers. The migrants that returned to their villages of origin took with them this pattern of thinking, relating skills to economic mobility.

At the same time, nearby, a large industrial complex was established in the 1960s which gave young people in the region access to factory employment. Management in the factory very early on realized they had to carry out a deliberate campaign in the region fostering cultural change to 'civilize' the people in the localities (cf. Arizpe, 1978). This means to turn peasants into factory workers, or to be more precise, to have young people adopt appropriate attitudes that would provide the rationale for them to accept the conditions of factory work. The campaign to foster cultural change was successful and the firm has brought in more than twenty million pesos annually to the area in wages. As a result, the whole life style of the area has changed. Among other things, school enrolment jumped in Toxi at the beginning of the 1970s, to such an extent that parents are now sending their children to the school in a nearby town, Ixtlahuaca, because the teachers are said to be better there.

In contrast, in Dotejiare, the only wage labour available is in agriculture. Opportunities in commerce or transport are monopolized by the Mazahua élite and a few mestizo families. Migration among the peasant households follows a strictly family and kin type of strategy. Land is still available, so that the younger generation can return to agricultural activities. As a result, school attendance has to be induced by the school principal and the *cacique*. Even families where children's labour is not needed send their children to school reluctantly. Dotejiare migrants in Mexico City have tended to keep within the protective 'shell' of Mazahua identity. Thus, ethnic ties are still more important than skills in ensuring economic survival of mobility.

GOVERNMENT POLICY AND RURAL CULTURE

The Mexican Government has not formulated a policy of cultural change since Vasconcelos and the 1930s, when it generally advocated the spread of literacy, a strong nationalism in the arts, and indianism. Subsequent régimes have taken a *laissez faire* attitude regarding culture, while attention has been focused almost exclusively on economic growth. At the local level, the result of this lack of policy can be seen in the drop of the quality of schooling and a feeling of aimlessness in Government projects directed at culture, such as the *Misiones Culturales*, (the social workers sent to the villages in the 1940s and 1950s to foster cultural change), and in a decline of cultural expression in the region as a whole. Documentary evidence indicates that at the beginning of the century, the larger towns in the region had some cultural life: local people wrote on its history, its landscapes, and the lives of those around them.

The political centralization of the new post-Revolutionary Government in the 1920s reflected itself in a marked cultural centralization in Mexico City. Interestingly, the novels, films, and mural art of the 1920s and 1930s still had their roots in rural society: they dealt with the *hacienda*, the Revolution, and the *charro*, the national figure. Songs and singers of rural origin were promoted, and films acted by Dolores del Río, María Felix, Jorge Negrete, and others, often under the masterful direction of Emilio 'Indio' Fernández, vividly portrayed the passion and violence of the countryside. In the 1950s, the revolt against an excessive nationalism in the arts and the rise of the urban middle classes shifted the pivot of Mexican culture to the city. Whereas before, in the 1940s, both the rural and the urban were reflected in films and novels, in the 1950s and 1960s, the urban has predominated to the almost total exclusion of the rural.

At the local level, this can be clearly traced. Up until the 1950s, the radio constituted the main cultural linkage to Mexico City. According to those who remember, programmes were good: the *locutores*—radio broadcasters—read poetry and commented on events; the *radionovelas*—serialized melodramas—were very popular. But songs and singers of an exclusive Mexican heritage formed the core of cultural transmission on the radio.

By the end of the 1950s, the quality of radio programmes declined as attention began to turn towards television in the city. Slowly, programmes on the radio became mostly music, much of it American, and news broadcasts became sensationalist. The favourite programme on the radio in the Mazahua region in the 1970s was an 8 a.m. news programme containing information about the latest murders in the vicinity.

As literacy, or rather, semiliteracy increased, the region was deluged by cheap comics and *fotonovelas* (melodramatic stories in the cartoon format incorporating a variety of subjects ranging from adventure to romance and passion).

Television made its appearance on a larger scale at the end of the 1960s. Few families have one, but they charge fifty cents to other people who want to watch it, and these are mostly children. Television programmes are in a majority serialized melodramas or *telenovelas*, and dubbed American televized series. Viewing time is very heavily loaded with commercials showing a 'modern' way of life pointing towards consumerism. In effect, what has happened is that the mass media are now in total control of cultural change in the region. But they beam out an almost exclusively urban culture. The effect of this 'modernization' has been a sense of alienation in the region: young people, particularly, despise the rural way of life and everything having to do with agriculture. As a result, cultural change is not shaping itself to fill the needs and configuration of a society that derives its livelihood from agriculture, but those of the urban one in Mexico City. The rural culture has been robbed of its dignity and its future: there is no place for it in the charter of development for the Mexican society.

CONCLUSIONS

Ethnic filiation has directly affected peasants in the Mazahua region by channelling forces that destroy the peasant economy, on the one hand, and that provide wage labour and allow capital accumulation, differentially among ethnic groups. The Mazahua indian group has had to bear the impact of the former, yet it has been barred from access to the latter. The volume and speed of 'passing' from the Mazahua ethnic group to the mestizo group was seen to be directly affected by economic opportunities and by Government policy favouring this process. In general terms, it was shown that the existence of an ethnic boundary channels the process of proletarianization so that the class line will again coincide with the ethnic line.

Analysis clearly indicated a dialectical relationship between economic mobility and ethnic filiation, both in the rural area, as well as in the city. In the latter, migrants depend on kin groups and the ethnic community to get jobs, housing, and financial help. If overall expansion of employment allows them mobility, they tend to leave the ethnic community and to change their cultural habits. If occupational mobility is not possible, then they tend to stay within the group and its economic specialization. However, reciprocally, those who, from the beginning, never venture outside the work and social area of the ethnic group have less possibilites of achieving occupational mobility. The availability of employment is, however, the necessary condition for this to be a relative disadvantage.

Mazahua migrants unable to achieve economic mobility in Mexico City have taken two strategies: either they retain their ethnic identity, enabling them to participate in their economic specialization and to receive collective support in crises, or, if their Mazahua ethnic ties are already too tenuous, they enter a newly created ethnic grouping vaguely rooted in the past, the *Concheros*.

Schools in the Mazahua region are not instrumental in an absolute sense in bringing about literacy and 'modern' attitudes, that is, behaviour appropriate to industrial work and consumerism. School attendance implies a rejection of the Mazahua identity, and the latter will only be abandoned if economic survival and mobility do not depend on belonging to the ethnic community. Only where opportunities for widespread wage employment exist, as in Santiago Toxi, or where constant migration links the village to the pattern of industrial employment, do parents become interested in increasing the level of schooling of their children.

The lack of a coherent, explicit cultural policy emanating from the Mexican Government after the decade of the 1940s has shifted control of cultural change in the country to the mass media, as shown by data from the region. Also, the rural content of Mexican culture portrayed in the cinema, the literature, and the arts until the 1950s has almost entirely disappeared. Mexican culture has become predominantly urban, middle class, and outward looking. At the local level, this has shown itself in the changeover from good-quality radio programmes to poor-quality comics, radio, and television programmes. In all of them, rural life is portrayed as backward and degrading, with no future of its own. As a result, an alienated culture is spreading in the region, more dependent than ever before on outside initiatives.

BIBLIOGRAPHY

Arizpe, L., 1978, *El Reto del Pluralismo Cultural* (Mexico: Instituto Nacional Indigenista).
Bonfil, G., 1972, 'El concepto de indio en America: una categoría de la situación colonial', *Anales de Antropología*, **IX**, 35–45.
Foster, G., 1960, *Culture and Conquest* (New York: Wenner-Gren Foundation for Anthropological Research).
Kemper, R., 1970, 'El estudio antropológico de la migración a las ciudades en América Latina', *América Indígena*, **XXX**, 3–20.
Lerner, D., 1958 *The Passing of Traditional Society* (Glencoe, N. Y.; Free Press).
Pitt-Rivers, J., 1973, 'Race in Latin America: the concept of "raza,"' *Archives of European Sociology*, **XIX**, 3–31.
Ribeiro, D., 1968, *The Civilizational Process* (Washington: Smithsonian Institute Press).
——— 1975, *Las Américas y la Civilización* (Mexico: Nuestro Tiempo).
Roberts, B. R., 1974, 'The interrelationships of city and provinces in Peru and Guatemala'; in Cornelius, W., and Trueblood, F. (Eds), *Latin American Urban Research*, Vol. 4, (Beverly Hills: Sage), pp. 207–36.
Singer, P., 1975, *La Economía Política de la Urbanización* (Mexico: Siglo XXI).
Stavenhagen, R., 1974, 'La sociedad plural én America Latina', *Dialogos*, **55**, 5–10.
Van den Berghe, P., 1974, 'Ethnic membership and cultural change in Guatemala'; in Heath, D. (Ed.), *Contemporary Cultures and Societies of Latin America*, (New York: Random House).
Wolf, E. R., 1959, *Sons of the Shaking Earth* (Chicago: University of Chicago Press)

Environment, Society, and Rural Change in Latin America
Edited by D. A. Preston
© 1980 John Wiley & Sons Ltd.

CHAPTER 9

Ethnicity in the Southern Peruvian Highlands

STEVEN S. WEBSTER

1. INTRODUCTION

In the recent, and only, sociological analysis of the Cuzco region of southern Peru, van den Berghe and Primov (1977, p. 259) baldly suggest that 'unless the ethnic factor is taken into account, Cuzco's great leap forward into the 20th century [out of the 18th century] may well turn out to be a throwback into the "socialist empire" of the 15th century'. They point out that, although agrarian reform since 1969 has substantially changed the key class factor of the means of production, oppression of the indian masses continues on an ethnic, if not a class, basis. That is to say, Peruvians tend to use cultural and linguistic criteria to entrench class inequalities. Although van den Berghe and Primov do not elaborate on a means of implementation, they recommend support of ethnic pluralism and local self-determination in Peruvian public policy, mentioning educational and linguistic reform as a means to countering ethnocentrism between the country's ethnic groups.

My thesis in this essay will only support an extend this conclusion, and probably make no advance on problems of its implementation. On the other hand, I am not so ready as van den Berghe and Primov to write off the agrarian reform as a failure, however hamstrung by bureaucratic involution and doctrinal exuberance it may still be. Although land and other capital distribution has been considerably reformed, it is far too soon to assume that the entire economic base of the class structure should by now have changed sufficiently to overcome class oppression. Nor am I so sure that a reversion to a 'socialist empire' or some form of centralized progressive régime would necessarily be a bad thing for Peru. At the very least, the populist ideology which sells the new régime is reaching the most withdrawn communities, promising official recognition of their worth. However dubiously these promises are received, this is nevertheless everywhere affecting community self-concept and initiative. As I will argue here, the social and cultural foundations are already well developed for these opportunities to express themselves in ethnic terms. The rising international consciousness of ethnicity, borne as pervasively into the highlands in mass media as anywhere else, further encourages this mode of expression. Regardless of whether or not the Peruvian reform is changing the class structure of the southern highlands, van den Berghe and Primov, and I agree that ethnicity is a far more important factor in that structure than most government planners and social scientists are ready to admit (also see van den Berghe (1974) and Webster (1970) for reviews of relevant literature).

My contribution will be to enlarge on van den Berghe and Primov's thesis, shifting it further towards 'the ethnic factor' than they themselves were able to do, and padding it out a bit with some pertinent case material. My perspective is based upon more intensive, if not such extensive, field experience, and benefits more from a view of the southern highlands system from the ethnic 'outside' of the class system, whereas van den Berghe and Primov's view is primarily from the urban inside of that system. We have both spent about the same time in the highlands (I in 1969–70; van den Berghe and Primov in 1972–73), although I returned for four months in 1977. They devoted most of their time and research to a suburban town outside of Cuzco, with thorough research excursions in the Department of Cuzco all along what they have viewed as a rural–urban continuum. On the other hand, I have devoted virtually all of my research to one rather isolated rural indian community, although I have travelled by foot, horse, and truck in rural areas of Cuzco and Ancash Departments and also lived for long periods in Cuzco itself, at the other end of the rural–urban continuum. Although my perspective may be biased by the anthropological enthusiasm for the apparently remote, strange, and traditional, theirs is probably biased by the sociologist's naive urbanity. And, although we both enjoyed the mixed anthropological blessing of the outsider's perspective on society, van den Berghe and Primov seemed to have become sufficiently integrated in the suburban sector of their continuum to have remained only dilettanti in its rural extremes. Or, to use what to my mind is a more apt metaphor for a plural society, they hardly viewed the coin from both sides. Consequently, although their analysis of departmental social organization is thorough (and almost unprecedented; I know of only one other such attempt for the Peruvian highlands: Bourricaud, 1967), their analysis of ethnicity, key to their entire thesis, remains pallid. Indeed, their notion of ethnicity in Cuzco Department is almost swallowed by that of class, from which they set out to distinguish it.

Before I review the major perspectives on class, ethnicity, and plural society in this area of Peru, I want to sketch the social organization of the particular community from which I have developed my own perspective on the question, paying particular attention to the political economy of ethnicity.

2. ASPECTS OF SOCIAL ORGANIZATION

Cheqec is one of several ethnic regions spread out along the *Ceja de la Montaña*, the high eastern flank of the Andean cordillera overlooking the Amazon basin, between Cuzco and Lake Titicaca. Other such ethnic regions in the *ceja*, intermontane valleys, and *puna* (the high Andean steppe) are variously called

Q'ero, Ch'ilka, Ch'uchu, Qolla, Qeshwa, and others. These terms appear on no map, but are part of the folk tradition of the area, and designate different regional sorts of Quechua-speaking people, perhaps originally tribal affiliations. The Cheqec, at least, consider each of the other ethnic groups as distinct kinds of people, although all of these including themselves are *runa* ('people'), as distinct from *misti*. *Misti* or mestizos are Peruvians generally distinguishable by their Western European cultural characteristics, and reciprocally they characterize *runa* as *indios, indigenas*, or with more derogatory ethnic terms. To a naive outsider, all *runa* would appear to be similarly poor peasants of similarly simple culture, struggling to derive a barren existence from the severe and supposedly inhospitable Andean fastnesses. To someone more familiar with the area (e.g. resident mestizos), each ethnic category is distinguished by its region of origin, one or a few traditional weaving motifs displayed on their mens' dress ponchos and womens' mantles, and characteristic personality stereotypes. To someone still more familiar with one or another ethnic region, certain distinctive cultural patterns such as fiestas, pilgrimages, political organization, or kinship customs are distinguishable, their reality resting as much in ideology as in behavioural differences. Of course, as one gets still closer to understanding and intimate involvement in specific situations, many of these supposed similarities and differences, the stuff of ethnicity, become vague in shifting perspectives, and all that is clearly left are individual people of awesome variability, more often challenging the ethnic paradigm which constrains them than conforming to it.

The ethnic region of Cheqec ranges along the eastern Andean flanks for about 150 km, and is itself comprised of about a dozen communities. These share the features of the general Cheqec culture but are distinct from one another through their history, fiesta, and political organization, relatively high rate of endogamy, and separate status under district or *hacienda* administration. The community with which I am most familiar, and which I will simply call Cheqec ('dispersed' or 'scattered'), is centrally located in the wider Cheqec region. It is composed of a rugged basin about 30 km wide in which five small rivers converge from separate valleys, descending from glaciers at 5000 m to subtropical forest at 2000 m above sea level. About 400 Cheqec people live in these valleys, in eleven small hamlets, usually located at about 4100 m for optimum pasturing of their herds of alpacas and llamas. These herds are the most important aspect of both their subsistence and trade economies. At intermediate altitudes (3300–4000 m), a wide variety of Andean tubers are raised which form the staple diet, and maize, as well as a few subtropical vegetables, are cultivated in the lower extremities of the community (2000 m).

Prima facie, the distinctive ethnicity of the community is based upon geographic integrity, a viable subsistence economy, social, political, and religious organization which is discontinuous with mestizo culture, and a mutual ideology of separateness from mestizo culture. I have described these aspects of cummunity organization in three other essays (Webster, 1973, 1975, 1977).

Briefly, a native form of camelid pastoralism, horticultural transhumance, and locally based pantheons of dieties and their native intermediaries are more or less common to all communities of the ethnic region, integrated closely with a vertically seriated ecological system. The community political organization is distinct from the dominant mestizo or State polity, based upon a system of status rank rather than stratification, and associated with distinctive criteria of wealth and prestige derived from religious fiesta, shamanism, and cultural brokership. The basis of social organization is a parallel kindred, related by current marriages and affinal statuses to other such kindreds. Marriage tends to be exclusive, resulting in about 80 per cent community endogamy and virtually complete endogamy within the ethnic region.

The Cheqec ethnic region is bounded to the south and west by a small but virtually impassable range of 5500 m peaks and glaciers, and to the north and east by precipitous descent into tropical jungle which, so far as I know, no Cheqec has ever ventured to penetrate below about 1800 m altitude. To both the southeast and north-west, or more directly along a route which flanks the mountain glaciers, lie major provincial market towns and roads, 60 to 100 km from the innermost community on mountain trails traversible only on foot or horseback. Although this strikes the outsider as stultifying isolation, it is a common situation for many of the Quechua communities in other elhnic regions of the southern Peruvian Andes as well. Even the Cheqec of the community I describe make several visits each year to one or another provincial market centre, and sometimes younger, more adventurous, family members will go as far as Cuzco, Sicuani, Quince Mil, or Puerto Maldonado, major departmental urban centres (see Figure 9.1). As far as I can determine, these sallies into the alien mestizo world are not the recent results of increased sophistication and acculturation, but have always occurred, at least within the memory of the oldest inhabitants, as a matter of economic course and social curiosity.[1] Similarly, itinerant buyers and vendors, both mestizo and from other ethnic groups, frequently visit the community on business at particular times of the year. However geographically remote and ethnically distinctive, the Cheqec of even the most interior community cannot be viewed as socially or culturally isolated. On the other hand, in the outermost Cheqec communities, much closer to the mestizo centres of communication, this cosmopolitan concourse has been more frequent and generally resulted in somewhat more economic and political dependency.

Vestiges of long-disused terraces and irrigation networks suggest that the Cheqec were probably more firmly subjugated to a colonial régime under the Incas than ever since. Probably sometime in the eighteenth century the entire region became an *encomienda* under the administration of a *corregidor* recently immigrated from Spain. By the end of the nineteenth century it was divided among his descendants into several *haciendas* congruent with the present separate communities of the region. Most of these *haciendas* were not directly administered, but rather their owners lived in the distant provincial centres, or

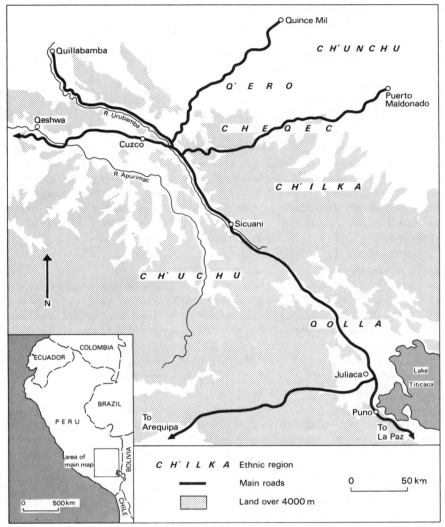

Figure 9.1 The Cuzco region of Peru

even in Cuzco or Lima, and administered their holdings indirectly. At least in the case of the innermost community of Cheqec, native leaders were appointed to administer the tasks of cultivation, harvest, and delivery of produce, because the inhabitants were sufficiently uncooperative with mestizo or outside administrators to render any more rigorous control counterproductive. Also, suggesting that the *hacienda* régime did not maintain a clearly upper hand over the inhabitants were its holdings in potato plots, which were not generally the largest nor most fertile, and its non-interference with alpaca and maize production, the more critical concerns of the native economy.

Although preoccupation with the exploitive aspects of highland *hacienda* régimes has resulted in contrary emphasis in the literature, I feel that such compromise accommodation between a remotely administered régime and native inhabitants of a *hacienda* has probably been frequent in this area of Peru (cf. Bertram, 1974; Martinez-Alier, 1973). In the case of Cheqec, a notoriously oppressive owner had acquired the *hacienda* by 1945, and, visiting it regularly and unpredictably from his residence 100 km away, demanding personal attendants in his household, and occasionally despatching the provincial police, he sought to exact subservience if not greater production. By 1964, this situation had sufficiently aggravated the Cheqec to produce some rebellious leaders, one of whom demanded a local school and was jailed by the owner for his efforts to organize it. The case came to the attention of influential liberals in Cuzco, one of which succeeded in pressuring the Government (then facing widespread peasant unrest) into expropriating the *hacienda* in favour of its inhabitants. Another Cheqec leader was instrumental in mediating this expropriation and organizing the long-term repayment of a Government loan. Since this time, the community has been legally independent and self-administering, its leaders creatively confronting the increasingly complex needs of cultural brokership with mestizo society. By 1977, it would have fully paid off its bank loan for expropriation were it not for the advice of their mentor in Cuzco (who continues to counsel them) that the eventual upshot of the subsequent national agrarian reform may involve the writing off of all such debts due to the insolvency of many of the resulting cooperative farms.

It is significant that despite the rebellious indignation of the Cheqec which precipitated the expropriation, a large faction of the community favoured continuation of the *hacienda* régime, even under the very unpleasant owner of that time. Evidently for many, the advantages of a patron in the provincial centre, able to act as both representative and buffer between the Cheqec and the agencies of mestizo society, outweighed the indignities of the *hacienda* régime, compromised as it was through decades of mutual accommodation. The confrontation between community factions for expropriation and for accommodation of the *hacienda* régime also precipitated two innovative leaders for the latter cause. These became involved in regional peasant movements and attempted to implement in the community the notions of syndicates and unified demands as an alternative to expropriation. Even in 1977, thirteen years after expropriation, the old *hacienda* house in the distant provincial capital continued to be a meeting-place and source of temporary employment, and the widow of the previous owner continued to be a patroness or advisor to many Cheqec visiting the town or passing through. Probably much like the former system, the relationship continues to be a matter of enlightened mutual opportunism rather than simple dependency and domination.

From the perspective of this example, one may view the relationship between some highland *hacienda* administrations and the inhabitants of their land as a political and economic symbiosis characteristic of plural society. This is obscured by presumptions of domination, class oppression, and long-suffering peasant passivity. Furthermore, these latter presumptions may be fundamentally ethnocentric, demeaning the capacity of the peasant or *runa* to maintain a viable option in his ethnicity despite potential or actual oppression.[2]

Now I will briefly describe several of the social institutions through which the Cheqec, although ethnically distinctive, are also intensively interrelated with mestizo society.

As was the case with their accommodation of the *hacienda* régime, the Cheqec have to a considerable degree been able to dictate the form of their relationship to mestizo society, adapting its importunities as well as its opportunities to their steadily changing way of life. This ethnic initiative is clearest in regard to their political and religious relationships with mestizo society. It is also apparent in economy, schooling, and other sources of ideological influence, although in these regards mestizo domination in the highlands in generally overwhelming.

The Cheqec, as is generally the case with highland ethnic regions, are very minor participants in the regional market economy, producing and consuming little that passes through it. Generally, in such regions, there are no markets or stores except those in district capitals and other mestizo settlements. Exchange among *runa* is based largely on local or long-distance individual trade relations. The total cash value of market and extracommunity trade exchange for the average Cheqec family does not usually exceed the equivalent of US $30 per year. This is generally in the form of alpaca and sheep wool, some maize and peppers, and perhaps a few sheep or cattle sold on the hoof, traded for such necessities as coca leaf, salt, and small quantities of lower-altitude grains, sugar, kerosene, dye, homespun fabric, and cooking utensils. This external exchange is usually carried out through the mediation of petty mestizo merchants who either come through the community or maintain shops in the market towns. Usually as *compadres* in ritual co-parenthood and widely known throughout Cheqec, these intermediaries are as much controlled in a highly formalized social relationship as they are controllers of the economic exchange. In this way, the Cheqec, if astute, can minimize their losses and perhaps occasionally maximize advantages (cf. Flores, 1974, 1976). Some special exchanges, such as maize for dried meat and wool for homespun, are traditionally carried out with other *runa* ethnic groups whose intinerant merchants visit annually. These seem to have social rather than economic justification, because the Cheqec can make these items more cheaply than they can trade for them. Only occasionally have relatively expensive purchases such as manufactured clothing, lanterns, Primuses, foodgrinders, or (recently) radios been made by a few families. There is considerable stigma upon purchase of such luxuries, and high value is placed upon self-sufficiency and hard work to derive from the diverse ecological zones of the community all that they have traditionally produced. On the other hand, surplus or specialized production is looked upon with suspicion. The individuals who defy these norms are either relatively wealthy and prestigious, poor, or young. If such defiance continues, increasing ostracism, gossip, and accusations force either departure from the community or exceptionally generous support of community feasts. The image which appears to justify this behaviour is not of a 'limited good' (Foster, 1965), but rather of *runa* ethnicity. The result is not a homogenization of peasant poverty but rather the development of considerable differences of wealth and prestige which are ethnically defined (Webster, 1975).

Another important aspect of external economic exchange is the community's cooperative production of a surplus of fresh and frost-dehydrated potatoes, derived from the former *hacienda* plots in the traditional manner of work-bees. Since expropriation, this has been devoted to repayment of the bank loan for purchase of title to their lands. Several of the wealthier families were able to pay off their share of the purchase price of 300 000 *soles* (about US $7500 at 1964 exchange rates) in a lump sum equivalent of US $150, and the remainder (poorer or more sceptical) contribute several days of cooperative labour each year to produce a harvest which the community sells for the equivalent of US $300–400. It is interesting to note that this special-purpose surplus was adapted from *hacienda* production to *hacienda* purchase, without much change in its form, and probably remains comparable to earlier forms of tribute production for other colonial régimes such as the Inca empire or colonial *encomienda*. In 1977, the bank loan would soon be paid off, and the Cheqec were considering whether to devote this annual surplus to pesticides and animal remedies or to some improvement in their school. In this way, they continue the gradual opportunistic redirection of what began as their exploitation towards new ends integral to their ethnicity.

Finally, among their primary modes of economic interrelationship with mestizo society is emigrant labour. Of about 87 adult males in 60 domestic groups, only about ten annually leave the community for periods of work up to a few weeks. This is usually unskilled *hacienda* labour within the ethnic region or, more rarely, in urban jobs outside the ethnic region, such as making adobes or carrying burdens. I know of only four of five individuals who have left for more than a few months, and in the period between 1970 and 1977 only two of these had left and not returned to the community. Objectives of emigrant labour are usually limited, such as earning money for a specific purpose or, in the case of a few youths, to experience more fully the alluring mestizo world. It is also usually undertaken by poorer members of the community who, for one reason or another, have given up one or more of the traditional aspects of subsistence economy such as the raising of maize or alpacas, and consequently sacrificed fuller economic independence. The Cheqec have a mildly derogatory term (*purih*—'walker') for such emigrant workers, viewing them as somewhat irresponsible and profligate. On the other hand, the greater familiarity of these men with mestizo ways and agencies often results in their use in making sales or purchases, and delivering messages, bank deposits or responses and requests to officials on behalf of the community.

Similar to external economic exchange, formal primary schooling in the mestizo tradition is a pervasive influence which both disintegrates and reintegrates Cheqec ethnicity. The Cheqec themselves have from the beginning been fully aware of the ambiguous value of schooling to their community. In about 1960, the *runa* innovator whom I mentioned earlier succeeded in convincing the community that schooling could broaden the opportunities of their children. It was hoped that through literacy and ability to speak Spanish the school would furnish the community with individuals more capable of dealing with mestizo agencies in matters of land reform, police, agricultural assistance, and similar pressing problems. However, hopes became disenchanted that schooling could fulfil these ideal goals and interfere no further than was desired in the Cheqec way of life. Attendance has usually been less than 30 per cent of school-age children. Parents found that although children acquired some abstract familiarization with the mestizo world, what little Spanish was learned was neither used nor retained. Furthermore, use of this language, along with moralistic overtones in the teaching of most subjects, served mainly to derogate *runa* culture and extol that of mestizos. This insidious influence was tolerated as long as the teacher himself (always mestizo) did not intrude in community affairs unless asked, and as long as education might produce leaders who could act as cultural brokers for the community. Nevertheless, it is primarily the poorer families, those whose viable options within the community's way of life are most limited, who most consistently send their children to attend the two or three years of schooling offered. Meanwhile, the Cheqec have learned to utilize the more cooperative teachers as cultural brokers, and discourage or virtually ostracize those who do not align themselves with community goals. Teachers write messages, explain community points of view to outsiders, and occasionally arbitrate conflicts which threaten to invite the unwanted attention of provincial officials.

Other significant sources of ideological influences, perhaps ultimately even more important factors in emerging or declining Cheqec ethnicity, are soccer and radio. Although the characteristically mestizo frenzy of these institutions is recently penetrating even the most remote plateaus and valleys of the Andes, they carry the stigma of the *cholo*, a kind of *misti* proletarian jet-setter. Younger Cheqec defiantly save for a radio and set aside Sundays for soccer practice, staccato commercials shatter the tranquillity of more traditional customary conversations and joking in the early hours of dawn, and spiked shoes and gaudy socks clash with the more traditional colours of dress ponchos. But these remain concerns of the young, and give way to more important preoccupations if one aspires to respected adulthood in the community. Meanwhile, however, all within earshot of the radio are deluged by whole constellations of mestizo values dressed in attractive commercialism and presented in native Quechua. Those involved with soccer competition in other communities (although so far only other Cheqec groups) become familiar with interesting ideas and eligible females with which they would otherwise have had little contact. I think there is some probability that these influences will draw away an increasing number of youths to emigrant labour. In 1977, I encountered a few more youths than in 1970 who were outspokenly rebellious and disparaging of the community. On the other hand, such influences are not qualitatively different from those which have produced a steady trickle of emigrants for several generations, as long as the Cheqec can recall, and the great majority of these adventurers have returned to settle down as good citizens. Those few who have not returned leave no impression on the community, and those who do return add as much to the scepticism regarding opportunities in the mestizo world as they do to the sense of cultural deprivation which all ethnic groups experience, insofar as they are aware of other ways of life.

Even the overwhelming institutions of mestizo economy and schooling can be significantly transformed under the resistance, selectivity, and innovation of community ethnic integrity. This process of cultural renovation is clearer in Cheqec politics and religion, where several centuries of mestizo policy directed at domination and assimilation have resulted instead in new syntheses so integral to Cheqec culture that they are hardly recognizable as exogenous. On the other hand, the contemporary political and religious culture is misinterpreted, indeed demeaned, if the continual effects of mestizo influence are not appreciated. For instance, although the Catholic rites of passage in baptism and marriage have been fully integrated with indigenous rites of passage and the political organization associated with them (Webster, 1975), the intercession of a mestizo priest is considered necessary to obtain the sacrament which consummates these rites. Although the Cheqec may not see a priest in more than a decade (priests rarely visit the community), they are sure to secure these services when they do. Conversely, a highly independent system of dispute arbitration within the community is devoted to resolution of conflicts before they exceed community boundaries precisely to avoid the intervention of mestizo authorities. Similarly, an important factor in prestige and community political organization is cultural brokership, the ability to mediate between Cheqec and mestizo social institutions without compromising the former. Cheqec ethnicity responds to mestizo culture, whether through its integration or its exclusion.

Spanish post-Conquest colonial policy, like that of the Incas before them, established local control and exploitation under the sponsorship of religion. Although it is likely there were Incaic or other antecedents, the colonial system of native *varayoq* ('those with the staff of office'—the mayor and his assistants) was probably established by the Viceroy Toledo in 1572, closely supervised by the priesthood and associated with religious festival *cargos* (Vasquez, 1963).

ETHNICITY IN THE SOUTHERN PERUVIAN HIGHLANDS

Probably as a consequence of this colonial institution, one finds contemporary indigenous society with systems of local leadership which arise from rites of passage, responsibility in religious feasts, and special relationships with supernatural powers. Regardless of its colonial origins, the result is often distinctively ethnic. Subsequent political offices established in these communities under *hacienda* or State dominion, such as *mandones, mayordomos, gobernadores, teniente gobernadores*, and *personeros*, have similarly often been integrated in various ways with this 'indigenous' power structure. Like the feasts and rituals, nominally of Catholic origin, which mark the Cheqec's assumption of new political status, these rôles constitute an archaeology of ethnic dialectics.

The *varayoq*, also called in Quechua the *kamachikuq* ('those who cause things to be done'), are a hierarchy of several offices through which most male Cheqec ascend by late middle age as official community authorities also responsible for the larger feasts of the year. These are calendrical feasts of the Catholic Church which have been adapted to the celebration of indigenous rituals such as renewal of leadership, new clothing, and fertility of the herds (Webster, 1973). The *varayoq* are themselves subordinate to a diffuse and unofficial oligarchy of fifteen to twenty prestigious elders of the community, the most successful alumni of this system. The *alcalde* ('mayor'), whose power is in any case limited in a community suspicious of authority, is often seen conferring with one or another elder of this group prior to and during feasts, prior to annual appointment of the new *varayoq*, over community disputes, and at times of other crises such as need for some action with or response to mestizo agencies. A leader's rise within the several offices of the *varayoq* proceeds under individual sponsorship by one or more of these elders, who may loan to their novice ritual paraphernalia such as the staff of office or conchshell trumpet, and act as divinatory intermediary to spirits and dieties.

The *varayoq* may have been established when the community elders were suspect in the sixteenth-century Church inquisition against idolatry. In Cheqec, the latter have since become a covert oligarchy which effectively controls what was originally an unwelcome imposition of mestizo influence. The ancient State ties of the *varayoq* are still reflected in their annual investiture before the provincial subprefect, although this is mere formality and always, so far as I know, acquiesces to the will of the community. The later *hacienda* office of *mandón* wielded considerable power under the old régime, and was probably intended by the *hacienda* owner to counter the control of the *varayoq* in much the same way that Toledo had probably intended the *varayoq* to overcome the influence of the elders. The independence of the modern *varayoq* is attested by the *hacienda* owners' unsuccessful attempt in the late 1950s to appoint one or more mayors not of the community's own choosing. The process of this co-optation of external control may be reflected in the development of the *gobernador* and *personero* rôles, established in Cheqec as extensions of State authority in the 1950s. Through 1970, these offices were generally occupied by individuals with little or no influence in the community, devoted to minor responsibilities unrelated to their official purpose, and functioned as buffers between provincial authority and the actual community leadership. Increasingly since 1970, individuals with considerable influence have been appointed to these offices, but their actual function is now cultural brokership on behalf of the community. The Cheqec are wary of the power of this rôle, and so far it remains subordinate to the *varayoq* and elders.

3. CLASS, ETHNICITY, AND PLURAL SOCIETY

I have reviewed aspects of social organization and leadership in one particular community representative of an ethnic region of Cuzco Department, emphasizing its modes of integration with mestizo society. I have argued that this integration, rather than eroding ethnicity and resulting in inclusion in the urban-based mestizo class structure, has instead promoted an on-going cultural renovation and ethnic vitality which continues to supersede class relations. Rather than passive subordination as peasants in the dominant class system, the Cheqec maintain an ethnic initiative in a symbiotic political economy, earlier with a *hacienda* régime, and now with a State régime. Even in economic exchange and schooling, spheres of influence usually overwhelmingly dominated by mestizo values, the Cheqec are able largely to dictate the form of their relationship to mestizo society through a combination of self-sufficiency, resistance, selectivity, social stigma, guile, and opportunism, guided by the initiative of innovative leaders. Most clearly in political and religious aspects of their culture, a dialectic between the *de jure* dominance of mestizo political economy and *de facto* initiative of the community organization has produced a stratification of co-opted institutions. These are continually reintegrated in a whole which cannot be viewed as autochthonous but is nevertheless distinctively Cheqec. Indeed, to consider contemporary Cheqec culture as autochthonous in any sense demeans it, whereas its ethnic integrity is most apparent if viewed as a creative response to pre-colonial, colonial, and modern mestizo cultural influences.

It now remains to review alternative perspectives on class, ethnicity, and plural society in this area of the Andes, comparing them with my own. Van den Berghe has already distinguished several divergent tendencies in this analysis (van den Berghe, 1974; van den Berghe and Primov, 1977), but I must go somewhat further in order to distinguish my own view from that of van den Berghe and Primov.

A prevailing theoretical perspective among contemporary Peruvianists (e.g. Fuenzalida, 1970; Greaves, 1972) derives from the influential Latin American analyses of Andre Gunder Frank and Rodolfo Stavenhagen. This perspective has also been instrumental in reformist Peruvian Government policy since the late 1960s. Although it is generally termed 'sociology of dependence,'

I will call this perspective 'class reductionism'. Briefly, it is argued that all major social and cultural differences in modern Peruvian society derive from differential access to the means of production. Consequently, the society can be explained in terms of classes which subscribe to a common scheme of values but unequally achieve them. Invocation of ethnic differences is seen as gratuitous. At least with regard to Cuzco Department of southern Peru, van den Berghe and Primov's examination of ethnicity and class stratification suggests that this approach is excessively simplistic. They argue convincingly that ethnic factors are always pervasive and sometimes critical in the explanation of social inequality, that neither ethnicity nor class are reducible to one another, and that neither can be studied without regard to the other. Ethnicity is constituted by cultural aspects of race and language, and the values which underwrite the key institutions of family, kinship, religion, and the political economy. On the other hand, the social relations of differential power, influence, and production which actually organize a society (whether or not this is congruent with its cultural values), constitute classes. So, for instance, van den Berghe and Primov (1977, p. 259) argue that in the case of Peru, even though relations of production have been radically changed by the agrarian and industrial reform, class inequality is entrenched by ethnic differences such as language, dress, and locality of residence.

Marxist preconceptions aside, I feel that the class reductionist perspective has simply not become aware of 'the other Peru'. The sociological literature of the south Central Andes adequately surveys rural mestizo communities, and the peasantry directly affiliated with these communities and consequently with the urban-based class system. But there is a long-standing sociological ignorance of the more remote rural populations, and consequently of what may be termed 'native' Quechua communities which form a large part of them (cf. Webster, 1970). The majority of the very dense population of the Central Andean plateau is rural, so even if all provincial and district centres of population are viewed as integrated in the urban-based class system, large numbers of people in villages, hamlets, and scattered farms (*parcialidades, villorios, anexos,* and *estancias*) may nevertheless be viewed more as members of regional ethnic groups than of this class system. Although the Cheqec ethnic region is probably unique in specific cultural and ecological terms, as the mestizo bias of ethnography is corrected it increasingly appears that ethnic distinctiveness itself is not unusual in the south Central Andes. Such ethnic groups may be scattered in even smaller regional populations that that of the Cheqec region (which probably contains about 2500 people) but they are far more extensive than urban centres and their total number is at least a considerable minority. From my own experience, it appears that in addition to the Cheqec region, several of the adjacent *hacienda* territories may be characterized as groups with varying degrees of ethnic initiative within the regional class system. Recent reports suggest ethnic regions in Chia (Ollachea province of Puno Department; Christinat, 1972) and other communities of eastern Chancis, Quispicanchis province of Cuzco (Gow and Condori, 1976),

Alcavitoria (Chumbivilcas province of Cuzco; Custred, 1973), the districts of Paratia (Puno Department; Flores, 1968), Kaykay (Paucartambo province of Cuzco; Flores, 1974), and generally throughout the herding communities of the *puna* plateau including northern Bolivia and the Departments of Apurimac, Huancavalica, and Ayacucho as well as Puno and Cuzco (Flores, 1975; Custred, 1973).

Ethnographic ignorance and ethnocentrism probably underlies the illusion of total class stratification in the Andes, obscuring plural society simply because it is not apparent in the urban centres. Especially in the ideological climate of social revolution fostered in Peru since the late 1960s, the class reductionist perspective has dominated publications by Peruvian social scientists. But urban centres and the system of class stratification based in them are only one side of the Andean coin. The other side is a vast, heterogeneously ethnic, rural presence to which the colonial ideology and political economy of mestizo culture has always reacted. On the other hand, this heterogeneous ethnic presence can in no way be reduced to the anthropological romanticisms of autochthony or 'cultural dualism', as much red herrings as class reductionism.[3] 'The other Peru' is the other side of the coin of Andean social organization in the sense that it is no more analysable apart from particular regional histories and class systems than they are analysable apart from it.

Bertram (1974) and Martinez-Alier (1973) have advanced class analyses of the Peruvian peasantry in a critique of 'the sociology of dependence'. They argue convincingly that the peasantry have generally not passively accepted their exploitation, but have struck a balance of advantages and disadvantages in an authentic class struggle with landowners. Consequently, they have often remained conservatively sceptical in the face of capitalist rationalization and even agrarian reform. Whereas this perspective acknowledges a far more active rôle for *runa* than the more simplistic class reductionism, I still feel it fails to recognize the uniquely ethnic form that such confrontation has often taken.

Van den Berghe and Primov's assessment of social inequality in Cuzco Department now offers an alternative view to class reductionism, putting in clear and comprehensive terms the case for significant cultural differences. These are viewed along a rural–urban continuum of ideal types, reflecting a gradual dominance of class over ethnic factors. However, they do not oversimplify, and their examination of class organization in the primary institutions of highland society is, if anything, more painstaking than their parallel concern with ethnic factors. They conclude that a 'continuum of ethnicity' informs the social stratification of the Department, with ethnic and paternalistic relationships predominating at the rural extreme, while class and competitive relationships are most salient at the urban extreme (van den Berghe and Primov, 1977, pp. 4, 123, 132, 252, 258). The 'classic plural society' characteristic of Guatemala and Chiapas is found only in communities of the rural extreme. Here, the ethnic line is far more sharply drawn due to the absense of the middle ranges of the continuum, leaving only a local mestizo oligarchy and a paternalized mass of mildly hostile but impassive and subservient indians (van den Berghe and Primov 1977, p. 123). Both are bereft of leadership by emigration of innovative and socially mobile individuals to urban centres, leaving local ethnic lines intact and developing no local middle class.

On the other hand, according to van den Berghe and Primov (1977), while ethnicity is salient in communities of the rural extreme, there is 'a near identity of class and ethnic status', with indians constituting a 'bottom ethno-class' of full-time peasants (pp. 118–9, 255–6). Because 'the Peruvian stratification system might be described as one of open ethnic lines and fairly rigid class lines ... paradoxically it is the permeability of ethnic boundaries that keeps those who remain indians at the very bottom' (p. 141). Elsewhere in the rural–urban continuum, 'depluralisation' (pp. 257–8) or an 'amalgamation' (pp. 121–3) or blending of cultures is far advanced, so that, on the whole, the highlands cannot be seen as a plural society in any narrow sense (p. 257). Dissident ethnic separatism is nearly impossible because open ethnic 'passing' and a pervasive class ideology co-opts potential indian leadership into the mestizo class system (p. 260). Unlike the 'classic' plural societies of Quetzaltenango (Guatemala) and Otavalo (Ecuador), where indians can rise in class status while remaining ethnically indians, upward class mobility in Cuzco inevitably involves change in ethnic status as well (p. 255).

This brief summary does not do justice to van den Berghe's and Primov's careful presentation. Their work is the most coherent and exhaustive analysis of highland mestizo society at the macro-regional level, and unique in its systematic application of a full battery of social stratification concepts. However, although they have broken new ground in their consistent attention to ethnic factors, I feel that, contrary to their goal, they have equivocated them with, rather than distinguished them from, class factors. From an effort to show that neither ethnicity nor class are reducible to one another, one would hope for clearer leads than the 'near identity of class and ethnic status' (p. 255), the 'ethnic cum class scale' (p. 122), and the 'near perfect correlation between the status of indian and [that of] peasant' (p. 259).

From my perspective, van den Berghe and Primov's conclusions are enlightening in their application to the urban-based class system including 'trunk-line' departmental, provincial, and distinct capitals, and furthermore to smaller localized centres of rural population in which a minority of mestizos maintain prevailing influence. But, again, 'the other Peru', so critical to an understanding of the ideology and political economy of ethnicity, has been largely left out of their picture. The ideal-type communities of Tocra and Q'ero might have been more fully segments of ethnic regions than of the highland class system, but one cannot assess this from the information presented. Judging from my own experience, it is likely that, in their brief examinations of these communities, the researchers were unable to penetrate the conventional front–rôle of lower-class servility which the Quechua native finds useful to exploit the outsider's paternalism, defend against threatening demands, and put off further enquiry. If one cannot penetrate this front, then the conclusion of 'a near identity of class and ethnic status' is likely to follow. With the Cheqec, I knew I was gaining understanding when I was accepted by many as harmless and useless, and tendered a rudeness reserved for those who might know too much. As van den Berghe and Primov appreciated, the *runa* or Indian changes his demeanour as easily as he changes his clothes to fit the ethnic occasion, both inside and outside his community. Given this long-established mode of defensive opportunism, I am not surprised that many ethnic regions are made partly invisible, often taken by outsiders simply as communities of impassive and toiling peasants.

I also think that van den Berghe and Primov (1977, pp. 123, 129) have given short shrift to the notion of plural society, begging the key question of the political economy in rural populations of indians and mestizos. Van den Berghe has elsewhere clarified the notion of plural society by predicating it on the integration of constituent ethnic groups through political and economic institutions dominated by one of these groups (Colby and van den Berghe, 1969). Unless we have reified institutions in our notion of a plural society, it goes without saying that, insofar as constituent ethnic groups are culturally different, the political economy common to the whole will not be homogeneously monocultural. But van den Berghe and Primov furnish few details on the nature or extent of political and economic integration between indians and mestizos in the most rural areas of Cuzco Department, other than an assumption of dominance and exploitation by the latter and envious subserviance by the former. Although they grant that ethnicity is 'salient' over class relations in such contexts, they beg the question of the form and degree of political and economic domination by mestizos. In some respects, they seem themselves to have adopted the assumptions of the class reductionist perspective they criticize, or at least to have been prone to the assumption of some 'inexorable and unilinear assimilation process' which has misled much analysis of ethnicity (Nagata, 1974). If there are sectors of rural society where such domination is weak or merely *de jure*, as is the case with Cheqec, then one cannot simply assume that all ethnic relations are subordinated within the class system. Such an assumption can misled enquiry, and obscure, for instance, the ethnic initiative and interplay between diverse levels of economy, polity, and ritual that I have described for Cheqec. Again, this dialectic reflects the history of a shifting balance of power between the community and the dominant sector, and the continual emergence of Cheqec ethnicity.

If we do not assume the *de facto* domination of class over ethnic relations, then neither can it be assumed without further evidence that indian leadership is sapped by the class system, that indian mobility necessarily implies change of ethnic status, nor that those who remain indians are necessarily kept 'at the very bottom'. Of the Cheqec leaders whom I mentioned, only one has left the community, while the others find challenge and success within the community and its ethnic tradition, and are dubious about whether the other individual's settlement outside the region has resulted in an increase of his relative social status. Van den Berghe and Primov grant that an indian can enjoy high status within his community, but seem not to take this possibility seriously as an ethnic option relatively independent of the class system. Elsewhere, I have described the elaborate system of social mobility active within the Cheqec ethnic option (Webster, 1975), and although I know a few Cheqec who have rejected it (and left the community or accepted low status within it), I know many more who pursue this challenge with all the fervour and satisfaction of a *cholo* scrambling up the mestizo class ladder. Nor can one indulge a patronizing attitude towards this alternative system of social rank when confronted by one of its *grandees*, a *runa* whose demeanour is awesome, whose glance is arrogant and defiant, whose influence spans several communities, and whose wealth in herds and patronage could buy out a mestizo *lumpenproletariat* several times over. It is possible for a naive outsider to regard or address such a *runa* as 'at the very bottom' (mestizos generally assume this with an indian, all of whom tend to appear the same to them), but it is perceived by insiders as a ludicrous charade and a confirmation of the mestizo's ignorant conceits.

Van den Berghe and Primov point out that the indian 'middle class' of Quetzaltenango, Guatemala, and Otavalo, Ecuador, are exceptions to the rule in Latin American plural societies. From what I have seen of La Paz, Bolivia, sectors of the local Quechua and Aymara ethnic groups there may be other exceptions to this rule. Furthermore, as one might suspect, given such geographically separated examples, these exceptions may represent ethnic tendencies in these plural societies, prompted by international factors such as decolonialization or its attendant minority ideologies. With regard to the contemporary Cuzco area, I suggest that the intra-ethnic mobility of indians within their community does not leave the mestizo class system unaffected, any more than mestizo class values leave particular *runa* systems of rank unaffected. When a rich and respected *runa* in his own community enters the provincial capital for the annual festival, mounted grandly with layers of rich ponchos and embroidery, mestizos are halted in their tracks, although their awe is soon modulated to contempt for this upstart *indio*. To cite another example, the mestizos of Ocongate, Canchis province, have boycotted certain prestigious annual religious *cargos* (fiesta steward), previously hotly contested among them, because indians of the Lauramarca region are carrying them off with more flair and resources. This region is one of the least unsuccessful cooperatives early established under the agrarian reform, and I would not assume that this bid for influence is simply the pretensions of newly successful *cholos* within the local mestizo class system. The reaction of the mestizo establishment suggests rather that the threat is from the outside, and that the new pretenders have retained their indian ethnic identity. The phenomenon of the 'rich indian', and his effects on local mestizo social organization, has also been perceptively reported by Flores for the Indian villages surrounding Kaykay, a mestizo town of Paucartambo province (in van den Berghe, 1974). Here, impoverished mestizos, however disdainful of the local indians, will nevertheless do field labour for them as *compadres* in order to benefit in trade from their horticultural wealth. To labour for another carries a strong stigma of subservience in the Andes, although in this case *compadrazgo* appears to rationalize what would otherwise be a status inversion. These examples suggest that the differences between ethnicity in Cuzco and in the 'classic plural societies' and 'middle-class indians' of Quetzaltenango and Otavalo are not so great as may be supposed.

4. CONCLUSIONS

In conclusion, I would emphasize with van den Berghe and Primov that ethnicity is a far more important factor in highland social organization than contemporary analysis has been inclined to admit. On the other hand, I go further than they do to argue that, in many respects, Peru is a 'classic' plural society, in which cultural divergence in many ethnic regions cannot be properly understood if viewed as subservient to class relationships. Authentic and viable ethnic alternatives besides the dominant mestizo culture continue in the highlands, and its diversity can be explained in no other way. Reduction of these complexities to class relationships or to a continuum model of amalgamation or assimilation invite misdirection of Government policy which will lead to further frustration of programmes of reform, and even to reactionary repression which subverts sincere intentions of social progress. The paternalism and bureaucratic involution which characterizes the policies and practice of the revolutionary Government since 1968 is now apparent (Harding, 1975; Bourque and Palmer, 1975; Handleman, 1974). Martinez-Alier (1973), Conlin (1974), Bertram (1974), and Gow and Condori (1976) argue that, given the limitations of technology and capital, the programmes of reform have frequently frustrated the interests of the peasants. They continue a passive resistance under bureaucratic exploitation which has not essentially changed their relationship to the means of production. Ironically, the reform appears to be impotent before the intransigent capitalist initiative of the peasants who, for instance, continue to give preference to their own plots and herds while devoting only token effort to communal cultivation, stock, and wage labour. Judging from my acquaintance with Cheqec leaders and ethnic strategy, I conclude that this behaviour reflects shrewd initiatives based upon social, ecological, and technological realities, not ignorant peasant conservatism. Given these realities, the best and perhaps only thing that agrarian reform can do is to give the lead to the divergent but progressive local interests of ethnic regions and *cholo* entrepreneurs. These can be supported through redistribution of land and technical assistance, but should not, perhaps increasingly cannot, be stifled by bureaucratic controls. If socialist ideals are feasible, they will be developed by granting self-determination of integral communities, i.e. to those which in particular regional histories have become solidary ethnically or in class structure. Policy must humbly adapt to the plural reality of Peruvian society. Ironically, like the smug deliberations of social scientists and development theorists, the policy of the Government is not likely to make much difference in the long run, confined largely to the rôle of 'tension management' (Bourque and Palmer, 1975) between the interests of various factions it cannot even claim to have created.

NOTES

[1] Thomas Zuidema told me in 1971 that in investigation of old documents he encountered references to the Cheqec reporting that they themselves were itinerant merchants in colonial times. This could refer to other communities of the ethnic region, or simply imply that they traded periodically over long distances the way many contemporary ethnic groups still do.

[2] This reassessment might also be applicable to some independant highland communities, which through loss of their land have become captive labour-pools for nearby *haciendas*. Of course, this interpretation should not be taken to condone such unjust distribution of land, but only to clarify the nature of class conflict in Andean society and to explain some of the frustration of the current agrarian reform in Peru.

[3] My own rather perfunctory effort in 1971 to criticize the class reductionist analysis in the Peruvian Andes was rightfully disparaged by van den Berghe and Primov as an anthropological romantic's defence of his 'cultural isolate'. I hope I have rectified that bias here.

REFERENCES

Bertram, I. G., 1974, 'New thinking on the Peruvian highland peasantry', *Pacific Viewpoint*, September, 89–110.

Bourque, S., and Palmer, D. S., 1975, 'Transforming the rural sector'; in Lowenthal A., (Ed.), *The Peruvian Experiment* (Princeton, N. J.: Princeton University Press), pp. 179–219.

Bourricaud, F., 1967, *Cambios en Puno*, Ediciones Especiales 48, Instituto Indigenista Interamericano, Mexico D. F.

Christinat, J., 1972, 'La mortalidad en Chia'; in *Allpanchis Phuturinqa* (Cuzco: Instituto de Pastoral Andina).

Colby, B., And van den Berghe, P. 1969, *Ixil Country; A Plural Society in Highland Guatemala* (Berkeley: University of California Press).

Conlin, S., 1974, 'Participation versus expertise'; in van den Berghe, P. (Ed.), 1974, pp. 151–66.

Custred, G., 1973, 'Puna zones of the south central Andes'; Paper presented in the *Symposium on Cultural Adaptations to Mountain Ecosystems* at the *Meeting of the American Anthropological Association*, 1973.

Flores, O. J., 1968, *Los Pastores de Paratia*, Serie Anthropologia Social, No. 10, Instituto Indigenista Interamericano, Mexico (Published in English, 1979, as *Pastoralists of the Andes; the Alpaca herders of Paratia*, Philadelphia: Institute for Studies of Human Issues).

—— 1974, 'Mistis and indians: their relations in a micro-region of Cuzco'; in van den Berghe, P. (Ed.), 1974, pp. 183–92.
—— 1975, 'Sociedad y cultura en la puna alta de Los Andes', *América Indígena*, **XXXV**, No. 2, 298–319.
—— 1976, 'El likira, intermediario ambulante en la cordillera de Canchis'; in *Antropología Andina*, Vols. 1 and 2, (Cuzco: Centro de Estudios Andinos).
Foster, G. M., 1965, Peasant Society and the image of limited good, *American Anthropologist*, **67**, No. 2, pp. 293–315.
Fuenzalida, F., 1970, 'Poder, raza, y etnía en el Peru contemporaneo'; in Matos, J. (Ed.), *El Indio y el Poder*, Serie Perú problema No. 4, Instituto Estudios Peruanos, Lima.
Gow, R., and Condori, B., 1976, *Kay Pacha* (Cuzco: Estudios Rurales Andinos).
Greaves, T. C., 1972, 'Pursuing cultural pluralism in the Andes', *Plural Societies*, **3**, Summer 33–49.
Handleman, H., 1974, *Struggle in the Andes* (Austin: University of Texas Press), p. 303.
Harding, C., 1975, 'Land reform and social conflict'; in Lowenthal, A. (Ed.), *The Peruvian Experiment* (Princeton, N. J.: Princeton University Press), pp. 220–54.
Martinez-Alier, J., 1973, *Los Huacchilleros del Peru* (Paris/Lima: Instituto de Estudios Peruanos–Ruedo Iberico).
Nagata, J. A., 1974, 'What is a Malay?—situational selection of ethnic identity in a plural society', *American Ethnologist*, **1**, No. 2, 331–50.
van den Berghe, P. (Ed.), 1974, Class and ethnicity in the Andes', *International Journal of Comparative Sociology*, **XV**, 3–4.
van den Berghe, P., and Primov, G., 1977, *Inequality in the Peruvian Andes; Class and Ethnicity in Cuzco* (Columbia, Mo.: University of Missouri Press).
Vasquez, M. C., 1963, 'Autoridades de una hacienda andina Peruana', *Perú Indígena*, **X**, No. 24/25, 24–36.
Webster, S., 1970, 'The contemporary Quechua indigenous culture of highland Peru; an annotated bibliography', *Behaviour Science Notes*, **5**, No. 2, 71–97; **5**, No. 3, 213–47.
—— 1971, 'Social organisation of the accommodated tribal society in highland Peru'; Paper read at the *70th Annual Meeting of the American Anthropological Association*, Seattle, 1970.
—— 1973, 'Native pastoralism in the south central Andes', *Ethnology*, **12**, April, 115–33.
—— 1975, 'Factores de la jerarquía social en una comunidad native Quechua'; in Bolton, R., and Forman, S. (Eds), *Conflicto e integración en los Andes, Estudios Andinos* **4**, **IV**, No. 2, 131–59.
—— 1977, 'Kinship and affinity in a native Quechua community'; in Bolton, R., Meyer, E. (Eds), *Andean Kinship and Marriage*, Special Publication No. 7 of the American Anthropological Association, pp. 28–42.

Rural Emigration

Environment, Society, and Rural Change in Latin America
Edited by D. A. Preston
© 1980 John Wiley & Sons Ltd.

CHAPTER 10

Migration and population imbalance in the settlement hierarchy of Argentina

RICHARD W. WILKIE

The rapid urbanization process and the resulting transformation of the settlement landscape and hierarchy are major developments affecting almost all countries in Latin America. Of the twenty major Republics, Argentina has become the most urban, and this has been primarily the result of internal migration since World War II. To understand potential changes in the rural sector of Latin America during the next quarter of the twentieth century, it is important to examine closely just how this urban migration process took place and to explore the implications for both rural Argentina, in particular, and rural Latin America in general.

While, in many ways, Argentina is unique, much of Latin America appears to be going through a very similar process. The rapid change in both the rural and urban sectors of Argentina has been paralleled, for example, by those in the state of São Paulo, Brazil. São Paulo, with a total population three-quarters that of Argentina, has had a changing urban–rural hierarchy since World War II that is virtually identical to that for Argentina, only ten years later. Much of southern South America, especially Uruguay, Chile, and parts of southern Brazil, have similar patterns of migration and urbanization.

What, then, is the situation in Argentina with regard to an expanding and urbanizing population? While it is the eighth largest country in the world in physical size, it has (*a*) a relatively small population of 25 million inhabitants, (*b*) a low population growth rate averaging 1.4 per cent per year between 1970 and 1975, (*c*) one of the lowest population densities in Latin America (8.6 per km^2 in 1972), (*d*) a highly educated and skilled population, and (*e*) a relatively large proportion of land in agriculture (62.3 per cent).

While Argentina has few problems associated with rapid population growth and density, the country historically has had an unbalanced urban–rural population distribution. As late as 1947, the vast majority of Argentines were split between two extremes of lifestyle, with nearly one-third dispersed over the landscape and another third living in metropolitan Buenos Aires. The nation lacked a developed sector of middle-sized cities, towns, and villages which would provide an adequate flow of goods, services, and information between levels in the urban hierarchy. By the 1970s, internal migration and continued industrialization in the largest metropolitan cities has reduced the dispersed population to about 15 per cent of the nation's total, while nearly half the population reside in the five largest metropolitan cities—Buenos Aires, Rosario, Córdoba, La Plata, and Mendoza. The population of Greater Buenos Aires is rapidly approaching 10 million, and it is the fifth largest city in the world.

To understand how the urbanization process led to this population imbalance, the settlement structure of Argentina over time and the movement of population between geographical regions will be examined. Additionally, a rural community will be studied to provide a more detailed view of rural change. Thus, the paper contains two major sections. The first part focuses on a national view of Argentina with three subsections: (1) a brief overview of recent changes in the agricultural population comparing Argentina to other Latin American Republics; (2) an examination of the urban–rural trends historically in Argentina; and (3) an analysis of internal migration based on ratios of in- to out-migrants by 1947, by 1960, and between 1965 and 1970. The second part is an examination of migration and settlement changes at the local level, in this case from a small village in Entre Ríos province. The volume of migration, the motivations for leaving, and the satisfaction with destination community size are examined. Using the findings at both the national and local level, the implications of rural depopulation in Argentina are explored.

THE STRUCTURE OF ARGENTINE POPULATION SETTLEMENT AND MOVEMENT

The argicultural population

While ranching and agriculture have dominated Argentina historically, the population has evolved into the most urban and least agricultural of the twenty Latin American Republics. By 1970, only 15 per cent of the population of Argentina was engaged in agriculture, the lowest in Latin America. Despite having the smallest proportion of population in agriculture in 1960 (20 per cent), this number decreased by 25 per cent in the next decade. Among Latin American Republics, only Venezuela lost a higher proportion of her population from agriculture between 1960 and 1970 (down 25.4 per cent). However, Venezuela had a larger population base in agriculture in 1960.

Argentina's reliance on agriculture has continued to decline dramatically both as a way of life and as a proportion of economic activity. By the early 1970s, it accounted for less than one-fifth (18 per cent) of the gross national product. The service sector led with 43 per cent of the GNP, followed by the secondary sector of manufacturing and construction with 39 per cent. Thus, Agrentina's economy is much more similar to those of Western Europe than to some of her Third World neighbours.

Urban and rural settlement trends in Argentina

While this shift from an agricultural to industrial economy is considered positive by most Argentines, population resettlement has not been spread evenly throughout the urban hierarchy. The predominant shift has been a sharp decline in the rural dispersed population and a sharp rise in large metropolitan centres of over half a million population, with an underdevelopment of intermediate-sized cities and towns. This section analyses how the urban and rural settlement hierarchy developed during the last 100 years using a typology of five settlement levels:

0 = Dispersed settlement: less than 200 inhabitants,
1 = Village: rural centres between 200 and 2000 inhabitants,
2 = Simple urban: centres between 2001 and 20 000 inhabitants,
3 = Complex urban: centres between 20 001 and 500 000 inhabitants,
4 = Metropolitan: centres over 500 000 inhabitants.

These five categories provide a more sensitive measure of the complex urbanization process than the simple dichotomy, urban–rural. By characterizing a region by the two categories holding the highest proportion of population, the evolution of the settlement landscape over time can be traced. Together, the two predominant levels generally account for between three-fifths and four-fifths of the population in the province or region, and thus quite accurately reflect the level of urbanization for the area at different points in time.

The logic behind these five settlement levels and the theory generated that stresses a balance between the levels is covered extensively in a study of Mexico (Wilkie, 1976a), and is not repeated here. There are a number of possible balanced distributions, and the more advanced countries have steplike progressions with somewhat larger proportions of population at each increasing size level of the urban—rural hierarchy. While other balanced situations are possible, it is essential that individual levels are not grossly underpopulated (or overpopulated) if the system is to function as an integrated unit.

In 1869, Argentina had a dispersed–simple urban (levels 0–2) settlement landscape. Two out of three people (67 per cent) lived dispersed on the rural landscape, not even clustering in small villages as large as 200 inhabitants. The level with the second highest percentage of population (17 per cent) was the simple urban level of communities between 2 000 and 20 000 inhabitants. Together these two settlement levels represented more than four-fifths of the population (84 per cent). While at that time the metropolitan level over 500 000 population did not exist, only 26 years later in 1895 those Argentines living in the metropolitan area of Buenos Aires had become the second most important group with 17 per cent of the nation's population. The dispersed level still dominated the settlement landscape with just under 60 per cent of the population, so by the end of the nineteenth century Argentina had become dispersed–metropolitan

Table 10.1 Proportions of Argentine population in five levels of the settlement hierarchy (percentages)

Year	Rural		Urban			Settlement landscape typology	Sum of two highest percentages
	Dispersed population (under 200)	Village (200–2000)	Simple urban (2001–20000)	Complex urban (20001–500000)	Metropolitan (over 500000)		
1869	66.9	4.5	16.7	12.4	0	0–2	83.6
1895	62.5	—	13.2	7.5	16.8	0–4	73.8[a]
1914	47.3	—	17.1	15.6	20.0	0–4	61.0[a]
1947	29.6	7.7	14.3	18.6	29.8	4–0	59.4
1960	17.5	8.7	16.0	17.6	40.2	4–3	57.8
1970	16.7	4.6	12.7	19.2[b]	46.8[b]	4–3	66.0

[a] Percentage estimated, since the censuses of 1895 and 1914 only listed rural population.
[b] La Plata and Mendoza were both counted as metropolitan, although both places were just under half a million population but were functionally in that category. The next largest complex urban centre has only 366 000 inhabitants.

(levels 0–4), with nearly three-quarters of the nation's population living in those two urban–rural levels (see Table 10.1).

While the urbanization process continued in Argentina during the first half of the twentieth century, it was not until 1947 that the population at the metropolitan level finally reached the same proportion as the dispersed level (30 per cent each). Thus, after mid-century the metropolitan level began to dominate the settlement hierarchy and, through concentration of population and economic power, to dominate the social and economic structure of the nation.

By 1960, the national settlement classification had changed from metropolitan–dispersed to metropolitan–complex urban due to the continued growth of metropolitan areas (from 30 to 40 per cent of the total) and a substantial decline in the dispersed population (from 30 to 18 per cent). While the percentage of population in complex urban centres of between 20 000 and 500 000 inhabitants remained about the same between 1947 and 1960, the dramatic drop in the dispersed population elevated this level to the second most important in the hierarchy. The rural exodus to various urban centres was so great between 1947 and 1960 that nearly half the land area of Argentina lost population (see Figure 10.1). Between 1960 and 1970, the number of areas losing population had decreased probably due to the already reduced population base living in the dispersed settlement and village levels, but about one-third of the rural interior continued to decline in population.

Greater Buenos Aires continued to grow rapidly and accounted for over half (54 per cent) of the population growth of the nation in the 1960s, thus demonstrating that migration flows to this primate city were still considerable. Simple urban centres, villages, and complex urban centres remained stable in proportion of total population, although both in- and out-migration were occurring at these levels. While stage migration took place, the majority of migrants in this time period went directly from dispersed areas to Buenos Aires. Germani (1961) found that three-quarters of the moves to a working-class district in Buenos Aires were 'direct moves' rather than 'step moves', although among later migrants this proportion had declined somewhat. Margulis (1968) found that four out of five moves (83 per cent) to Buenos Aires from a village in distant La Rioja province were 'direct moves'. Apparently, the economic and social opportunities of the intermediate-level cities were not enough to attract those migrants. Both studies, however, focused on migrants to the Buenos Aires metropolitan area and did not attempt to trace those migrants who had moved to other levels in the settlement hierarchy.

Between 1960 and 1970, metropolitan and complex urban sectors continued to increase (from 58 to 66 per cent), the population in simple urban centres fell by nearly a quarter (from 16 to 13 per cent), and the dispersed population dropped only slightly (from 18 to 17 per cent). Thus, at least until the latest census in 1970, the process of urbanization continued strongly. But population movements have become much more complex as direct migration to the largest cities has declined.

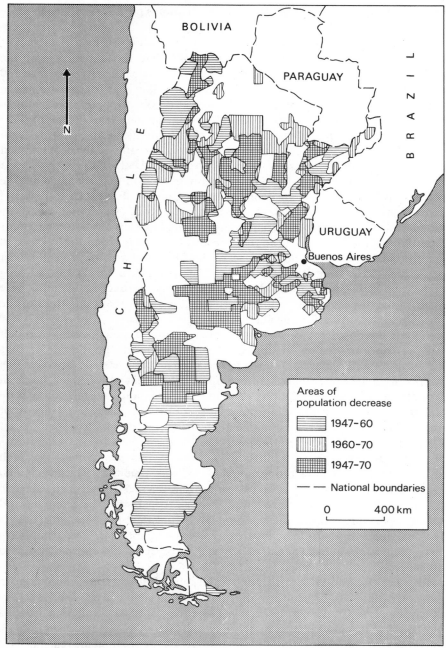

Figure 10.1 Depopulation in Argentina. Sources: Cole (1965) and Censo de Argentina 1970

In fact, one-third of all movements nationally between 1965 and 1970 were intermetropolitan, from the Capital Federal core of the metropolitan area to the Gran Buenos Aires suburbs and the reverse.

Internal migration in Argentina

Population shifts have occurred not only between levels in the settlement hierarchy but across provincial boundaries as well. National census data indicate only changes in provincial residence, but this provides a useful cross-sectional view of internal migration. In 1947 and 1960, the censuses asked province of birth and current province of residence, and thus a measure of lifetime migration is available. In the 1970 census, the question was where one had been living in 1965, thus revealing for the first time population shifts over a short period (five years) rather than over a lifetime. The data in all three time periods show only migration across provincial boundaries, so the amount of migration within provinces is unknown. In the case of Aldea San Francisco, at least one-third of all moves in this century were within the province of Entre Ríos, and thus would not be recorded in these national-level surveys. In larger provinces, such as Buenos Aires province, Santa Fé, and Córdoba, with more options in the settlement hierarchy for migrants, as many as half of the moves may be unrecorded. Still, a majority of all moves cross provincial boundaries in Argentina and reveal significant long-range shifts of migrants.

This section examines the relative health of regions in attracting, holding, or losing population by focusing on the ratio of out-migrants to in-migrants for each of the twenty-five provinces and geographical units in 1947, 1960, and 1970. These in- to out-migration ratios are illustrated for each period over cartogram base maps, where the size of the units are based on total population at that time rather than physical size. When examined periodically over a third of a century, these ratios provide a view of how internal migration changed the population distribution of Argentina.

The pattern of urban and primate city growth that characterizes Argentina in the post-1945 era was already evident in the migration trends by 1947 (see Figure 10.2). The Buenos Aires metropolitan area and Buenos Aires province had experienced considerably more in-migration than out-migration, as had Mendoza and virtually all of Patagonia. In addition, four provinces on the northern frontier were still growing through in-migration—Jujuy, Chaco, Formosa, and Misiones. Santa Fé and Córdoba in the central core and Salta in the north demonstrated stability through balanced ratios of migration, while nine provinces in the north-west and north-east had greater out-migration. Most notable in this latter group were Corrientes (6.3 out- to 1 in-migrant), Santiago del Estero (5.1 to 1), Entre Rios (4.9 to 1), Catamarca (4.7 to 1), La Rioja (4.2 to 1), and San Luis (3.8 to 1).

By 1960, the major trend continued to be the tremendous growth of Gran

164 ENVIRONMENT, SOCIETY, AND RURAL CHANGE IN LATIN AMERICA

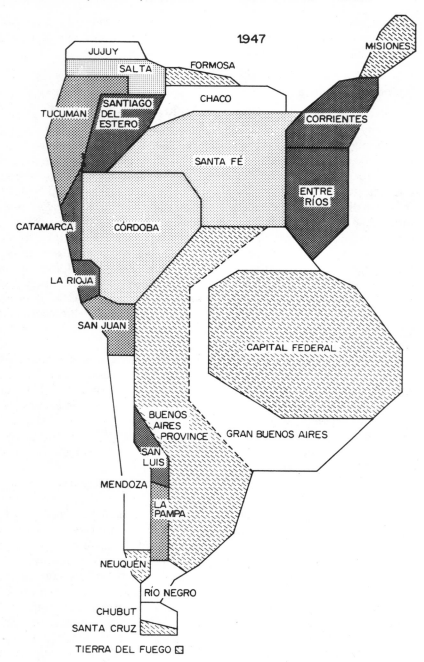

MIGRATION AND POPULATION IMBALANCE 165

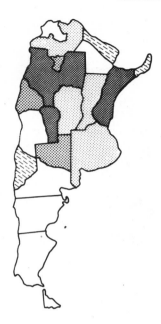

Argentina: 1947

	Type of migration: Out-	In-	Ratio: out- to in-
Central			
1. Capital Federal	452 634	694 984	0.65 to 1
2. Gran Buenos Aires	449 885	698 677	0.64 to 1
3. Buenos Aires Prov			
4. Santa Fé	270 789	252 957	1.07 to 1
5. Córdoba	206 419	198 107	1.04 to 1
Northeast			
6. Entre Ríos	187 544	38 182	4.91 to 1
7. Corrientes	180 064	28 643	6.29 to 1
8. Misiones	14 655	21 354	0.69 to 1
9. Chaco	29 715	142 826	0.21 to 1
10. Formosa	10 483	17 247	0.61 to 1
Northwest			
11. Jujuy	17 287	28 923	0.60 to 1
12. Salta	44 142	45 425	0.97 to 1
13. Tucumán	87 387	59 168	1.48 to 1
14. Santiago del Estero	145 371	28 224	5.15 to 1
15. Catamarca	59 971	12 761	4.70 to 1
16. La Rioja	45 299	10 880	4.16 to 1
West Central			
17. San Juan	37 590	21 063	1.78 to 1
18. San Luis	74 802	19 503	3.84 to 1
19. Mendoza	46 746	78 277	0.60 to 1
20. La Pampa	66 060	32 430	2.04 to 1
Patagonia (South)			
21. Neuquén	13 233	18 997	0.70 to 1
22. Río Negro	17 550	30 378	0.58 to 1
23. Chubut	12 640	24 055	0.53 to 1
24. Santa Cruz	7 724	8 970	0.86 to 1
25. Tierra del Fuego			

Cartogram size based on population in each province by 1947

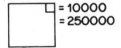
= 10000
= 250000

Ratios equal number of out-migrants to 1 in-migrant

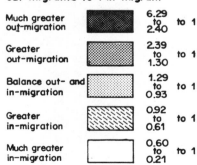

Much greater out-migration — 6.29 to 2.40 to 1
Greater out-migration — 2.39 to 1.30 to 1
Balance out- and in-migration — 1.29 to 0.93 to 1
Greater in-migration — 0.92 to 0.61 to 1
Much greater in-migration — 0.60 to 0.21 to 1

Figure 10.2 Migration in Argentina. Ratio of out- to in-migration by province, 1947, Source: ratios adapted from Lattes and Lattes (1969, cuadro 46); R. Wilkie, J. Marti, and M. Seawell.

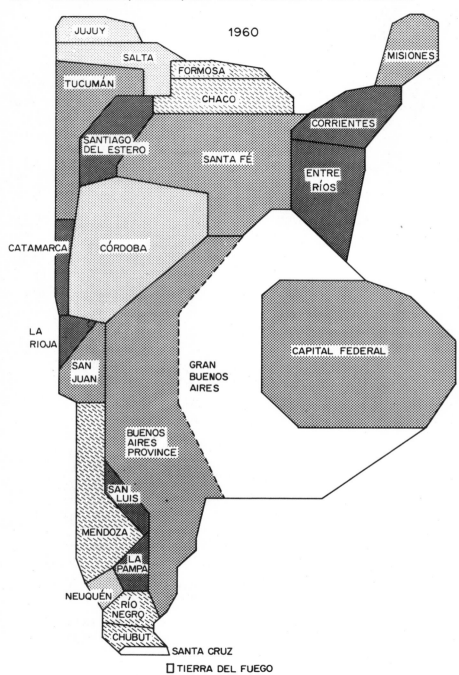

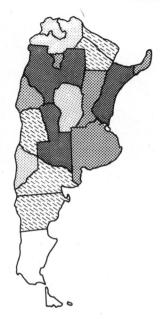

Argentina: 1960

	Type of migration: Out-	In-	Ratio: Out- to in-
Central			
1. Capital Federal	1 026 097	729 485	1.41 to 1
2. Gran Buenos Aires	124 576	2 057 673	0.06 to 1
3. Buenos Aires Prov.	815 428	384 046	2.12 to 1
4. Santa Fé	387 408	263 146	1.47 to 1
5. Córdoba	328 091	260 640	1.26 to 1
Northeast			
6. Entre Ríos	359 900	44 190	8.14 to 1
7. Corrientes	268 655	43 098	6.23 to 1
8. Misiones	48 633	27 430	1.77 to 1
9. Chaco	89 571	137 391	0.65 to 1
10. Formosa	22 235	25 193	0.88 to 1
Northwest			
11. Jujuy	37 948	39 572	0.96 to 1
12. Salta	64 366	63 983	1.01 to 1
13. Tucumán	147 490	96 662	1.53 to 1
14. Santiago del Estero	259 908	41 202	6.31 to 1
15. Catamarca	92 384	21 099	4.38 to 1
16. La Rioja	66 303	14 598	4.54 to 1
West Central			
17. San Juan	52 647	34 081	1.54 to 1
18. San Luis	109 203	25 899	4.22 to 1
19. Mendoza	78 751	120 420	0.65 to 1
20. La Pampa	104 185	32 868	3.17 to 1
Patagonia (South)			
21. Neuquén	25 378	20 236	1.25 to 1
22. Río Negro	37 191	44 533	0.84 to 1
23. Chubut	24 706	32 473	0.76 to 1
24. Santa Cruz / 25. Tierra del Fuego	7 810	18 946	0.41 to 1

Cartogram size based on population in each province by 1960

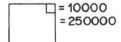
= 10000
= 250000

Ratios equal number of out-migrants to 1 in-migrant

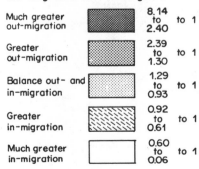

Much greater out-migration	8.14 to 2.40 to 1
Greater out-migration	2.39 to 1.30 to 1
Balance out- and in-migration	1.29 to 0.93 to 1
Greater in-migration	0.92 to 0.61 to 1
Much greater in-migration	0.60 to 0.06 to 1

Figure 10.3 Migration in Argentina. Ratio of out- to in-migration by province, 1960. Source: ratios adapted from Lattes and Lattes (1969, cuadro 46); R. Wilkie, J. Marti, and M. Seawell.

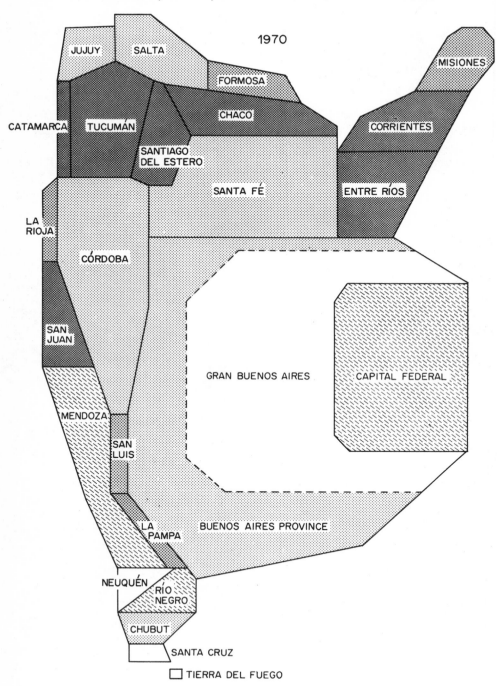

MIGRATION AND POPULATION IMBALANCE 169

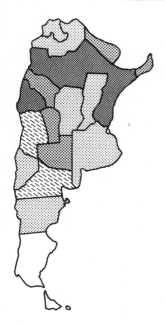

Cartogram size based on population in each province by 1970

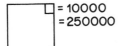
= 10000
= 250000

Argentina: 1965 to 1970

	Type of Migration: Out-	In-	Ratio: Out- to in-
Central			
1. Capital Federal	294 700	338 850	0.87 to 1
2. Gran Buenos Aires	175 800	801 900	0.22 to 1
3. Buenos Aires Prov.	272 150	210 400	1.29 to 1
4. Santa Fé	95 950	103 050	0.93 to 1
5. Córdoba	99 400	94 700	1.05 to 1
Northeast			
6. Entre Ríos	80 550	29 900	2.69 to 1
7. Corrientes	90 700	25 600	3.54 to 1
8. Misiones	45 150	26 450	1.71 to 1
9. Chaco	120 500	24 650	4.89 to 1
10. Formosa	27 150	17 450	1.56 to 1
Northwest			
11. Jujuy	29 000	25 650	1.13 to 1
12. Salta	39 450	32 800	1.20 to 1
13. Tucumán	89 600	25 550	3.50 to 1
14. Santiago del Estero	89 250	21 300	4.19 to 1
15. Catamarca	22 100	9 100	2.43 to 1
16. La Rioja	15 950	8 450	1.89 to 1
West Central			
17. San Juan	34 000	12 000	2.83 to 1
18. San Luis	19 700	11 600	1.70 to 1
19. Mendoza	42 400	49 700	0.85 to 1
20. La Pampa	16 050	8 250	1.95 to 1
Patagonia (South)			
21. Neuquén	14 750	25 350	0.58 to 1
22. Rio Negro	21 850	33 150	0.66 to 1
23. Chubut	16 350	16 550	0.99 to 1
24. Santa Cruz	9 550	16 600	0.58 to 1
25. Tierra del Fuego	2 200	3 900	0.56 to 1

Ratios equal number of out-migrants to 1 in-migrant

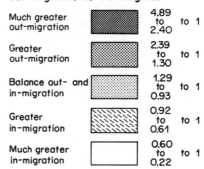

Much greater out-migration — 4.89 to 2.40 to 1

Greater out-migration — 2.39 to 1.30 to 1

Balance out- and in-migration — 1.29 to 0.93 to 1

Greater in-migration — 0.92 to 0.61 to 1

Much greater in-migration — 0.60 to 0.22 to 1

Figure 10.4 Migration in Argentina. Ratio of out- to in-migration by province, 1970, Source: Censo de Argentina 1970, R. Wilkie and M. Seawell.

Buenos Aires, which had more than two million in-migrants and increased its ratio to 17 in-migrants for every one out-migrant (see Figure 10.3). Other areas of in-migration continued to be Chaco and Formosa in the north, Mendoza in the west, and most of the Patagonian region in the south. The period from 1947 to 1960 experienced the greatest range of population ratios, from a high of more than eight out-migrants to one in-migrant in Entre Ríos to seventeen in-migrants for every one out-migrant in Gran Buenos Aires. Only Córdoba, Salta, Jujuy, and Neuquén had balanced migration ratios. Six areas with moderate and seven with considerable out-migration completely surrounded Gran Buenos Aires in the east, Córdoba in the central zone, and Mendoza in the west as urban growth centres. This was the time of the greatest movement of population out of rural areas.

Data collected in the national census of 1970 provide a different perspective on internal migration. For the first time, the question asked were one had been living five years previously, rather than where one had been born. The ratios indicate only the most recent movements between 1965 and 1970, and thus population shifts and the range of ratios were not as great (see Figures 10.4 and 10.5). Chaco province in the north had the highest net out-migration with nearly a 5 to 1 ratio, while the Gran Buenos Aires region had dropped to under a 5 to 1 ratio favouring in-migration. The 1965–70 pattern, however, remained similar to those of 1947 and 1960, except that the northern frontier had shifted from a destination area to either a stabilized (Salta and Jujuy) or out-migrant source area (Chaco, Formosa, and Misiones). Patagonia continued to attract in-migrants, except for Chubut, which showed a balance for the first time. Buenos Aires province and Santa Fé also developed balanced migration ratios after experiencing net losses by 1960, while the Federal Capital attracted enough new in-migrants to balance out the moderate losses experienced by 1960.

In examining the overall trends for the three time periods (see Figure 10.5), only three provinces had noticeably varied patterns. Chaco province in northern Argentina experienced the most extreme change, going from the most favourable position of all provinces by 1947 (5 in-migrants to 1 out-migrant) to the least favourable position in 1970 (1 in-migrant to 5 out-migrants). A similar, but less dramatic, pattern also occurred in Formosa and Misiones, neighbouring provinces of Chaco in northern Argentina. By 1947, these provinces were the principal areas of rapid frontier settlement. The ratio reversal in the 1960s suggests that the region did not live up to expectations, or perhaps attracted too many settlers for the infrastructure to support. Of the three provinces, only Misiones appears to be again reversing the pattern during the 1970s.

Nine provinces consistently had excess out-migration. Santiago del Estero and Corrientes (both bordering either Chaco or Misiones) had among the highest out-migration ratios, averaging approximately five out-migrants for every one in-migrant over all three time periods. Entre Ríos was third highest by 1947, first by 1960, and sixth between 1965 and 1970. Three provinces in the north-west and west—Catamarca, La Rioja, and San Luis—also have been consistently high

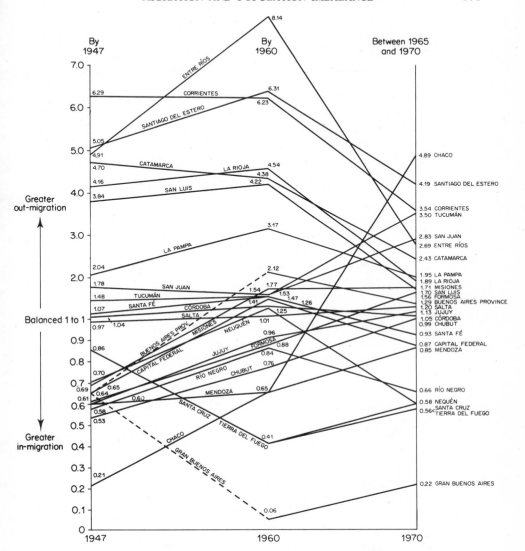

Figure 10.5 Ratio of out- to in-migration in Argentina, 1947–70. Data on Gran Buenos Aires and Buenos Aires Province were not kept separately in 1947, so both are connected by a dashed line. Source: data in figures 10.2, 10.3 and 10.4.

source areas, averaging between three and four out-migrants for each in-migrant. La Pampa and San Juan in the west central region had moderate population losses through out-migration. A bad three-year drought in La Pampa in the early 1950s accounts for the elevated out-migration ratio (3 to 1) found in 1960 compared to that in 1965–70 (2 to 1).

Five of the twenty-five geographical units experienced a generally balanced movement of migration over all three periods, and all except Salta were in the urban core of the country—Capital Federal, Buenos Aires province, Santa Fé, and Córdoba. Not surprisingly, Gran Buenos Aires was the unit with the highest in-migrant ratios over time, reaching a high by 1960 of nearly seventeen in-migrants to each out-migrant. Santa Cruz and Tierra del Fuego in the far south also had very high ratios favouring in-migration, but both began with small population bases. Two other provinces in Patagonia—Río Negro and Chubut—consistently had more in-migrants than out-migrants, as did Mendoza in the west central. Two provinces with moderately greater in-migration than out-migration were Jujuy in the far north and Neuquén to the south of Mendoza. In all, eight provinces have consistently attracted more migrants than they lost.

CASE STUDY CHANGING RURAL COMMUNITY AND THE OUT-MIGRATION PROCESS

While aggregate changes in the settlement hierarchy reveal the structure of migration and settlement change, they yield few insights into how these processes actually work at the local level. Aldea San Francisco, a village in Entre Ríos province, serves as an interesting case to view these processes. Proximate to Buenos Aires and Santa Fé provinces and with little industrialization, Entre Ríos has served as a prime source area of migration. By 1947, the outflow ratio of migrants from Entre Ríos was the third highest of any province (5 to 1). This ratio increased between 1947 and 1960 to the highest of any province (8 to 1) and then dropped to sixth place (2.7 to 1) between 1965 and 1970.

Aldea San Francisco is a community of over 200 inhabitants (315 in 1966, and 205 by 1977) located in the primarily agricultural western section of Entre Ríos bordering the Paraná River. Population decline in the last several decades has resulted primarily from out-migration, especially to the larger metropolitan areas such as Buenos Aires between the 1940s and 1960s and more recently to middle-level communities in Santa Fé province. The proximity of various-sized urban centres, combined with the isolation of the village—it is 5 km from a paved road and 7 km from piped water and a telephone—make it a typical village that is underdeveloped and losing population.

Aldea San Francisco in 1966 was in many respects the same village one might have seen in the 1870s when it was founded. Decendants of the six major extended families of Volga–Deutsch who migrated from the Volga region of Russia to settle in Aldea San Francisco were still the principal inhabitants 90 years later. A seventeenth-century German dialect maintained for 110 years in Russia was still as important as Spanish in the village. For some older women, it was the only functional language. A high intermarriage rate within the village and with other nearby Volga–Deutsch communities maintained the village's homogeneity and isolation. There was almost no agriculture mechanization;

horse-drawn carts were the principal vehical for farmwork and transportation. Occasionally, a Model-T Ford was seen passing on the dirt roads of the area. As late as the 1960s, Aldea San Francisco continued to be typical of similar villages throughout much of rural Argentina. Isolated ethnic and cultural communities, often attempting to maintain older traditions and lifestyles, were found throughout the north central, north-east, pampa, and Patagonian regions of the nation.

By 1974, during the relatively short span of eight years, Aldea San Francisco had moved decisively into the twentieth century. Advanced technology had directly reached about one-half the population of the village and indirectly touched the lives of all. Nearly 40 per cent of adult population (14 years and older) had access to a battery-operated television at home, and the television in the general store drew many of those without a set at home. Motorized vehicles also have been recently introduced. In 1966, 17 per cent of the adult population had access to a car or truck, but many of these vehicles were 1920s vintage and not functional for anything but short-distance travel. By 1974, nearly half the adult population were in a family which owned a car or truck, the majority purchased between 1972 and 1974. Further, half the adult population had access to a tractor, a proportion that has been steadily increasing since 1966 when there were only six old tractors in the village region. Change has again slowed, however, and between 1974 and 1978 these rates did not increase appreciably. So, while the village still does not have running water or paved roads, the isolation of the village has been significantly reduced by electricity, television, and automobiles; and machinery has begun to replace labour-intensive farm methods.

The purchases made since 1966 represent a sizable proportion of the yearly cash income of families. Gross incomes rose substantially during this period, even though differences in lifestyles of the more well-to-do and poor have increased, at least initially, with the introduction of more advanced technology (see Wilkie and Wilkie, 1975). For the top 20 per cent of families, mean annual gross income rose from $2212 to $12 350 (an increase of 560 per cent) during the eight-year period. For families without their own land or with only small plots, life continued to be a struggle. The gross annual family income of the poorest 20 per cent of families rose only from $123 in 1966 to $472 in 1974 (an increase of 380 per cent). In addition, between 1966 and 1974, nine families (nearly half the families in the poorest class) were forced to emigrate from the village after selling what possessions they did own.

Population turnover through migration in the village

Migration in to and out of Aldea San Francisco has occurred steadily during this century. Out-migration far exceeded in-migration and return-migration, as there was no way the village could absorb the high natural population increase resulting from the ten to twelve children typical of first-generation families. The

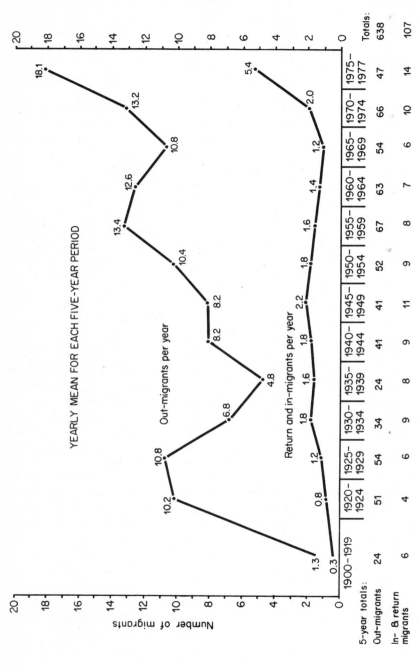

Figure 10.6 Migration volume in to and out of Aldea San Francisco, 1900–77. * The precise years are not known for twenty out-migrants prior to 1940.

volume of population movement in to and out of the village by five-year periods is shown in Figure 10.6. The highest out-migration occurred between 1950 and 1974, peaking in the late 1950s and again the 1970s. Since 1950, an average of between ten and eighteen migrants left the village population annually. In-migration has taken place consistently since the 1920s but in much smaller numbers. Most of these are females who married men from Aldea San Francisco, with the largest volume occurring since 1970. Return migration peaked between 1935 and 1949, following the world depression and World War II. In all, 598 permanent out-migrants are known to have left the village (nearly three times its present population) since its founding, while 65 in-migrants and 42 return migrants have moved back into the village after living elsewhere.

While the focus of this study is on the out-migration process, in-migrants and return migrants play important roles in the village, since they bring in knowledge and expertise from the outside. Return migrants, two-thirds of whom are male, provide a great source of technical and cultural change in the village.

While they account for less than one-fifth of the adult male population of Aldea San Francisco (19 per cent), they were named as first leader by two-thirds (67 per cent) of all villagers. Interestingly, most of these return-migrant leaders are from the middle rather than the upper class, the traditional source of leadership in the community. This new leadership provides a stimulus for modernization and change in the village (Wilkie and Wilkie, 1975).

Reasons for leaving Aldea San Francisco

Family and village records reveal that 408 migrants left the village since 1935. Of that total, about 70 per cent, or 278 migrants were located and interviewed in depth by 1974. Motivations for leaving Aldea San Francisco were examined by asking each migrant to score 49 items on a five-point scale ranging from a high of 'very much of a reason' to a low of 'no reason at all' for leaving the village. After scoring each of the 49 items, the migrant added any number of additional reasons he or she felt were important, and they were scaled. Once the list was completed, the interviewer read the items with the highest scores back to the respondent. Each migrant then determined which item from those with the highest scores was 'first in importance', which was 'second', and which ranked 'third'.

Responses to items ranged from 73.4 per cent saying that 'too little work in Aldea San Francisco' played at least some rôle in their decision to leave to only 1.2 per cent saying that 'debts and obligations in the village' played some rôle in the departure. The most frequently named items are listed in Table 10.2. Five of the six top-named items revolve around the job situation, although only two are clearly monetary. A clear 'pull' force from the outside appears to be operating, with a majority of migrants expressing desire for self-development and satisfaction with their occupation, work conditions, and activity settings. However, almost as frequently named are items related to family, social activity,

Table 10.2 Major Motivations For Out-Migration

Rank	Percentage saying it played some rôle	Reason for migration	Behavioural component	Push or pull
1	73.4	Too little work in Aldea San Francisco	Economic	Push
2	70.4	Better work conditions elsewhere	Environmental	Pull
3	68.3	To earn higher wages	Economic	Pull
4	63.0	To have work with individual development	Psychological	Pull
5	61.0	More to see and do elsewhere	Environmental	Pull
6	58.8	To have more satisfying work	Psychological	Pull
7	53.7	To move up the social ladder	Social	Pull
8	52.7	Bored with life in the village	Psychological	Push
9	52.3	To travel and see other places	Spatial	Pull
10	50.8	Family members lived in the new place	Social	Pull
11	50.4	To have job security	Economic	Pull
12	49.6	Preferred to live in 'modern' places	Cultural	Pull
13	48.4	Not enough land in Aldea San Francisco	Economic	Push
14	44.2	More events and activities elsewhere	Cultural	Pull
15	43.2	The village is too far from other places	Spatial	Push

Table 10.3 The Three Most Important Reasons For Migrating From Aldea San Francisco—264 migrants

		1st ranked	Percentage 1st	2nd ranked	3rd ranked	Total responses	Total percentage
1	Economic	103	39.5	85	78	266	35.0
2	Social	71	27.2	27	37	135	17.7
3	Environmental & spatial	29	11.2	52	58	139	18.2
4	Cultural	27	10.3	16	24	67	8.8
5	Psychological	19	7.3	51	32	102	13.4
6	Health & Diet	5	1.9	7	6	18	2.4
7	Religious	4	1.5	5	3	12	1.6
8	Political	3	1.1	10	1	14	1.8
9	Traditional	0	—	3	5	8	1.1
		261	100.0	256	244	761	100.0

travel, education, and distances to these things. 'Pull' forces from the urban areas outnumber 'push' factors, accounting for three-quarters of the top-named items compared to 60 per cent of the total list.

The analysis of the three most important reasons for leaving cited by the migrants indicates that (1) economic forces are the single most important motivation to move, and (2) other factors play a sizable rôle. Economic reasons were the most frequently named and ranked as the most important reason for leaving by about one-third of all migrants (see Table 10.3). Social reasons were the next most frequently named as first in importance, but dropped off considerably as second- and third-ranked reasons. Furthermore, when migrants were asked why they left at that particular time, as opposed to earlier or later, social factors outranked economic factors 35 per cent to 31 per cent. The importance of social motivation is further indicated by the fact that nearly three-quarters of all the reasons for migrating that were added to the list by the migrants themselves were social. These new responses included 'moved with family', 'to accompany spouse', 'family wanted me to join them', 'wife wanted to leave','too old to live alone', and others. Environmental and spatial reasons concerned with location and settings composed nearly one-fifth of all responses (18 per cent).

It is usually a combination of several key factors that leads to the decision to leave. There are always economic consequences of moves, and many migrants can most easily sum up a complex move by stating 'it was to get a better job'. Given a chance to evaluate other factors, the majority (61 per cent of first reasons and 65 per cent of the top three) ranked non-economic reasons as one or more of the three most important reasons for leaving the village. Three-quarters of the migrants said that to find a job played some rôle in their decision to migrate, but less than one-fifth (19 per cent) named it as the first-ranked reason. So, while the largest village subgroup thinks and acts primarily in economic terms, there are sizable migrant groups for whom family, environment and location, personal growth and satisfaction, or education is the principal reason for leaving. Clearly, various migrant groups seek a range of economic, social, environmental, and cultural conditions when they move. A balanced settlement hierarchy provides a greater range of choices and thus greater migrant satisfaction with life in all levels.

Satisfaction with position of community in the settlement hierarchy

The satisfaction of each migrant with their current location was measured in two ways: (1) satisfaction for the latest move, and (2) a comparison of their 'ideal' location in the settlement hierarchy with their actual location. Nine out of ten migrants from Aldea San Francisco said they were either 'very satisfied' (68 per cent) or 'satisfied' (20 per cent) with their latest move, although 45 per cent said they definitely would consider moving again. When asked to name the 'ideal'

Table 10.4 Current Location Compared with Perceived Ideal Location in the Urban Hierarchy—migrants from Aldea San Francisco, 1974 (percentages)

	MALES Ideally, would like to live in:					
Currently living in:	Village–dispersed	Simple urban	Complex urban	Metro-politan	Total (%)	N=
Dispersed (14.7%)	78	11	6	6	101	18
Village (10.7%)	38	38	16	8	100	13
Simple urban (33.6%)	32	14	22	32	100	41
Complex urban (8.2%)	40	10	20	30	100	10
Metropolitan (32.8%)	42	25	15	18	100	40
Total preferring each level (%)	43	20	16	21	100	
N=	53	24	20	25		122

10 males had no opinion
28% of males were satisfied with their current settlement level

	FEMALES Ideally, would like to live in:					
Currently living in:	Village–dispersed	Simple urban	Complex urban	Metro-politan	Total (%)	N=
Dispersed (9.0%)	18	27	46	9	100	11
Village (8.1%)	50	10	0	40	100	10
Simple urban (29.3%)	11	17	33	39	100	36
Complex urban (20.3%)	16	16	24	44	100	25
Metropolitan (33.3%)	20	2	17	61	100	41
Total preferring each level (%)	19	12	24	45	100	
N=	23	15	30			123

9 females had no opinion
36% of females were satisfied with their current settlement level.

location in the urban–rural hierarchy in which to live (see Table 10.4), only a little over one-third of females and one-quarter of males were currently residing at their 'ideal level' (Wilkie, 1980).

Male and female migrants from Aldea San Francisco are very different in what they want in the way of city size and urban complexity. Female migrants prefer less isolation and enjoy urban life more. Their highest rate of satisfaction is found in the metropolitan level, where 61 per cent want to remain. The metropolitan level has also the strongest attraction for migrant women living in both complex urban centres (44 per cent want to move to metropolitan areas) and simple urban centres (39 per cent want to go there), while 40 per cent of women living in villages also named it as 'ideal'. Together, seven out of ten female migrants from Aldea San Francisco opted for life in either metropolitan or complex urban centres when given the choice. Even women currently living in rural areas preferred the nucleated village to the open isolation of dispersed settlement, with 50 per cent of village women expressing satisfaction with their current location as opposed to only 18 per cent of women in dispersed areas.

Among male migrants from the village, on the other hand, there was a definite attraction for the rural areas (43 per cent) and, compared to females, much less for the two largest urban levels (37 per cent). In addition, eight out of ten males (78 per cent) who are already living in areas of dispersed population want to remain there, the only level in the settlement hierarchy other than village that received more than 20 per cent satisfaction among male migrants.

On one point male and female migrants show relative agreement—neither group currently living in simple urban or complex urban centres is particularly satisfied with life there. Only 14 per cent of males and 17 per cent of females would remain in simple urban centres between 2000 and 20 000 inhabitants if given the option, while for complex urban centres these percentages are 20 for males and 24 for females. Clearly these middle level centres fail to satisfy the needs of rural in-migrants. Part of the reason may be that few of the jobs, goods, capital, services, and related amenities that are concentrated in the metropolitan areas filter down to the middle-level centres. Some migrants to these middle-level cities stated that they have the urban problems without the compensation of real urban/cultural amenities, and they have lost the best of what the rural environment could offer as well.

CONCLUSIONS

As Argentina enters the last quarter of the twentieth century, it has evolved into a highly urbanized nation with only a minority (15 per cent) of her citizens engaged in agriculture. With some of the best land for agriculture in Latin America, low population density, and few problems related to population growth, Argentina nevertheless has a major population problem. With an overconcentration of jobs and services in the largest metropolitan centres, especially in the primate city of Gran Buenos Aires, middle-sized cities are not adequately developed as healthy regional centres where economic, social, and political power can filter down to the rural population. Due to a combination of 'pull' forces from these metropolitan centres and 'push' forces operating in the rural areas, a major shift of population from the rural to the urban areas occurred since the 1940s. Nine provinces, in a connected arch from La Pampa and San Luis in the west central, northwards through La Rioja, Catamarca, and Tucumán, and then eastwards through Santiago del Estero and Chaco to Corrientes and Entre Rios, provided a substantial proportion of out-migrants to these population flows. Encircled by this arch of considerable rural depopulation are the metropolitan dominated provinces of Buenos Aires, Santa Fé, and Córdoba. To the west of the arch is an area of growth in Mendoza and to the far north and far south are the settlement frontiers of the nation. Some of these frontier provinces continue to draw in-migrants (the Patagonian region in the south and Jujuy in the north-west), but others (Chaco, Formosa, and Misiones in the far north) have stopped attracting settlers and are now source areas of out-migration as well. Thus, the move to the Argentinian frontier settlement areas is currently only a trickle compared to the movements to urban areas. Land is no longer being given away, and considerable capital is needed to make a successful start into modern agriculture. Tenant occupied land dropped from 44 per cent in 1937 to only 16 per cent in 1960, while corporate, cooperative, State, and institutional landholdings rose from 18 to 34 per cent in the same period. Owner-occupied land also rose in those twenty-three years from 38 to 50 per cent of the landholdings. So, in spite of the popular conception of rural underpopulation in Argentina, it is clear that in mid-century there were too many inhabitants for the rural system to absorb adequately under existing land tenure and urban centralization practices.

A case study at the local level also contributed to our understanding of this rural depopulation process. First, it showed that economic change and modernization of farming and lifestyle at the local level was considerable between 1966 and 1974, even in a province like Entre Ríos, one of the three major source areas of out-migration. Secondly, total out-migration from the village this century numbers nearly three times the population living in the village in 1977. There is no way that this population could have been absorbed by the village, so for the most part migration has provided an important option for the villagers. Out-migration has reduced the population pressure on the land, allowed for increased landholding and, thus, farm mechanization, and has eased crowded housing conditions. The standard of living of those villagers profiting from these changes has significantly risen and almost all inhabitants have benefited in some way from the increasing modernization of the village. The influence of out-migrants and the important innovation and leadership of some return- and in-migrants has increased modernization in the village and provided access to many urban amenities. This circular flow of ideas, information, and migrants indicates that rural villages like Aldea San Francisco are strongly interwoven into the regional and national systems of modernization and change. And, for many of the out-migrants, particularly those from the middle class, life in the urban environment is a positive thing. The urban option represented the best chance of increasing their standard of living, and their hard work and achievement orientation has been a vital input to the urban sector of the nation. Thus, the volume of migration has increased steadily over the last thirty years, more often motivated by the 'pull' forces in the city than from 'push' forces in the village. Thirdly, while the most important cause of out-migration was economic, other factors—social, environmental, cultural, and psychological—together account for nearly two-thirds of the reasons people left the rural area. A complex combination of many forces lead to the movement out, and it varies by age, by sex, and by village subgroup. Finally, satisfaction of migrants with the five levels of the settlement hierarchy also varies. Female migrants overwhelmingly prefer the higher-order urban centres (or, if rural, the village cluster as opposed to dispersed settlement), while males prefer the two rural levels—especially dispersed settlement. Both males and females agree, however, that something is lacking in the simple urban and complex urban cities, as less than one-quarter of both groups are satisfied at this level.

The overriding problem of Argentina is that migrants have few options except to go to the largest metropolitan areas where jobs and services have been concentrated. One solution to this problem at the national level is to assure a strengthening of middle-level and lower-level regional centres that will create a balanced hierarchy of cities. The quality of rural life has been severely restricted by the absence of small urban centres capable of serving the dispersed population of the surrounding countryside with social meeting houses, movie theatres, schools, medical facilities, and, perhaps more importantly, small factory and service sector jobs that will hold people in an area. Without numerous focal points for economic and social activities dotting the landscape, the rural population is forced to maintain economic and social relationships with more distant regional capitals or even with Gran Buenos Aires. If the village and simple urban communities were strengthened, however, there would be a supporting infrastructure for a much larger number of people on the land.

With the rapid rise of the ownership of automobiles and trucks in the rural sector, the distances between the villages and the regional centres have been reduced, at least for the upper- and many middle-class rural families. However, automobile travel tends to bypass the lowest-order urban centres to reach higher-order urban centres with a greater range of services, thus weakening villages and simple urban centres more than they were previously. When this occurs, lower-class rural populations are isolated even more from economic and social options and are more likely to be forced out of the rural area in search of work. The population in villages of between 200 and 2000 inhabitants fell from 8.7 per cent of the national total in 1960 to only 4.6 per cent in 1970 (a drop in absolute terms of 667 000 inhabitants). Simple urban centres between 2000 and 20 000 population fell from 16 to 12.7 per cent during the same period (representing an absolute loss of 239 000 inhabitants). Since the nation's population grew by nearly 3.5 million between 1960 and 1970, and rural dispersed population in absolute terms increased by 409 000 to nearly four million population (despite a proportional decline of just under 1 per cent), these losses in already under-populated lower- and middle-order urban centres only added to the problems of maintaining an economically and socially healthy rural sector.

Plans to control and plan properly for growth of Gran Buenos Aires have been called for as early as the 1920s (Della Paolera, and Randle, 1977). Since the 1950s, the Argentine Government officially recognized the problem by issuing a series of laws, executive orders, and Ministry decrees aimed at limiting the growth of the Buenos Aires metropolitan area (Gambaccini, 1978), promoting industrial decentralization, and achieving a more balanced population distribution. These efforts, however, have been largely on paper. Despite this stated policy, the Capital Federal and the major urban provinces of Buenos Aires, Santa Fé, and Córdoba received between 1958 and 1969 nearly nine out of every ten (88 per cent) of all new federally approved economic investments. The gross industrial product for this central region reached nearly 90 per cent of the national total by 1968–69, with the Buenos Aires metropolitan area accounting for more than 50 per cent of the total (Rofman, 1977). Nor have attempts to open new areas of colonization been very successful. Between 1940 and 1968, the National Institute of Land Settlement and Tenure was able to settle only 9139 families on 898 652 hectares (J. Wilkie, 1974), and many of these have been from immigration rather than resettlement (Gambaccini, 1978).

A renewed effort by the Federal Government to lessen the concentration of economic activities and population may be making slight headway in the 1970s. The study of out-migration from Entre Ríos province in 1973–75 showed an increasing movement to intermediate-sized cities with small factories in Santa Fé and Entre Ríos provinces. Previous migration from the area between the late 1940s through the late 1960s had been dominated by movement to the Gran Buenos Aires metropolitan area. Urban violence, kidnapping of executives, and other factors such as noise and pollution involved in a declining quality of urban life as well as tax incentives for relocation have led to the movement of many small factories to interior towns. By 1977, factory relocation into the area south of Paraná, Entre Ríos, provided sufficient job opportunities to divert some migration from Gran Buenos Aires and medium-level cities in Santa Fé. Diamante, a simple urban centre of nearly 13 000 inhabitants in 1970 and the Departmental capital for Aldea San Francisco, recently received a leather processing and shoe factory that will employ 6000 workers by 1979. Whether this decentralization is sufficient to have a national effect and reverse the metropolitan concentration is as yet unknown. It may only mean that the expanding metropolitan landscape around Gran Buenos Aires that previously moved up the Paraná River to Rosario and the city of Santa Fé has finally crossed the Paraná River with the Santa Fé–Paraná tunnel and the new bridge at Zarate and will soon subsume the southern Entre Ríos landscape.

In conclusion, it is important to recognize that Argentina represents one extreme of the urban–rural continuum in Latin America. The shift from a rural to an urban economy and lifestyle that has taken place in Argentina will eventually be followed by many other countries of Latin America. If, in that process, the problems of primate city dominance are not considered, similar imbalances in the settlement hierarchy will result. There is no way that the rural sector of most Latin American countries can absorb their growing populations, especially after large-scale mechanized agriculture is introduced. Clearly, major rural-to-urban shifts of population will continue to take place in the entire region. If the settlement hierarchy develops in a balanced way, it will add to the options available to the rural population, the migrants, and to the satisfaction of those living in all size places.

ACKNOWLEDGMENTS

This reasearch was supported by a grant from the U.S. National Institute of Child Health and Human Development (project No. HDO 7397). In addition, the author would like to acknowledge data collection assistance from Alexander Paine and editional assistance from Jane R. Wilkie.

REFERENCES

Argentina, 1961, *Censo Nacional de 1960: Población*, Resultados Provisionles (Buenos Aires: Direccion Nacional de Estadistica y Censos).
—— 1963/69, *Censo Nacional de Población, 1960*, 9 Volumes.
—— 1971, *Censo Nacional de Población, Familias y Viviendas—1970*, Resultados Provisionales (Buenos Aires: Instituto Nacional de Estadistica y Censos).
Cole, J. P., 1965, *Latin America* (London: Butterworths).
Della Paolera, C. M., and Randle, P., 1977, *Buenos Aires y sus Problemas Urbanos* (Buenos Aires: OIKOS).
Gambaccini, P. J., 1978, *Population Movements and Government Response in Paraguay and Argentina: A Comparative Study*, Unpublished manuscript, Dept of Political Science, University of Connecticut.
Germani, G., 1961, 'Inquiry into the social effects of urbanization in a working-class sector of Greater Buenos Aires'; in Hauser, P. (Ed.), *Urbanization in Latin America* (New York: Columbia University Press), pp. 206–33.
Lattes, Z. L., Recchini de, and Lattes, A. E., 1969, *Migraciones en la Argentina* (Buenos Aires: Instituto Torcuato DiTella, Centro de Investigaciones Sociales).
Margulis, M., 1968, *Migración y Marginalidad en la Sociedad Argentina* (Buenos Aires: PAIDOS).
Rofman, A., 1977, 'La promoción industrial en la Argentina: Propuestas y resultados de los objectivos de descentralización regional', *Revista Paraguaya do Sociologia*.
Wilkie, J. R., and Wilkie, R. W., 1975, 'Migration and a rural community in transition: A case study in Argentina'; Paper presented at the *Conference on the Distribution of Population*, Belmont, Maryland, 29–31 January 1975, sponsored by the National Institute of Child Health and Human Development.
Wilkie, J. W., 1974, *Measuring Land Reform* (Los Angeles: UCLA Latin American Center).
Wilkie, R. W., 1974, 'The process method versus the hypothesis method: A nonlinear example of peasant spatial perception and behavior'; in Yeates, M. (Ed.), *Proceedings of the 1972 Meeting of the International Geographical Union Commission on Quantitative Geography* (Montreal and London: McGill–Queen's University Press), pp. 1–31.
—— 1976a, 'Urban growth and the transformation of the settlement landscape of Mexico, 1910–1970'; in Wilkie, J., Meyer, M., and Monzon de Wilkie, E. (Eds),

Contemporary Mexico: Papers of the *IV International Congress of Mexican History* (Berkeley: University of California Press), pp. 99–135.
—1980, 'Environmental perception and migration behavior: A case study in rural Argentina'; in Thomas, R. N.and Hunter J. M. (Eds.), *Internal Migration Systems in the Developing World* (Boston: G. K. Hall).

Environment, Society, and Rural Change in Latin America
Edited by D. A. Preston
© 1980 John Wiley & Sons Ltd.

CHAPTER 11

Change and rural emigration in central Mexico

HÉLÉNE RIVIÈRE D'ARC

Rural change and emigration has given rise to numerous studies by ethnologists, as well as by geographers and economists. Consequently, it is not intended to present new material in this paper but rather to portray the most general characteristics of emigration from the central plateau of Mexico. Rural emigration, whether temporary or permanent, is part of the process of change; it is both a result of change and at the same time it gives rise to further change.

The central plateau of Mexico is typical of those rural areas in Latin America that were much changed by settlement during the Colonial period and are densely populated. Its agrarian structures are similar to those of Spain, and the methods of cultivation and stock raising were those learned from the Spanish. The urban and communications networks that had been adequately set up in former times helped the integration into the market economy of this vast region which is at the heart of Mexico and the centre of diffusion of all types of activities.

However, the homogeneity of the central plateau contrasts with the periphery, with the poorer areas, or with the most traditional of the sparsely populated indigenous areas such as are found in the north, in the isthmus of Tehuantepec, which is now experiencing accelerated modernization, as well as in Oaxaca, Chiapas and Guerrero, where different local situations still hold strong and resist homogenization. It conceals a whole range of historical nuances which have been more or less strongly imprinted by settlement and by agrarian reform. Geographically, the homogeneity of the central plateau is illusory, for it contains many distinct ecological areas. Likewise, socially, there are different forms of stratification. Economically, there occurs more or less large-scale integration into the national market depending upon ease of communications and proximity of large centres of consumption, etc. And, finally, ethnologically, there is greater or lesser resistance to integration from place to place according to the level of internal cohesion of the different dominant groups (indian, western, or white). In other words, on the Mexican plateau there are *ejidos* (collective farms), small- and middle-scale farmers, some agricultural labourers, and communities that have been more or less affected by change, as well as the more or less powerful

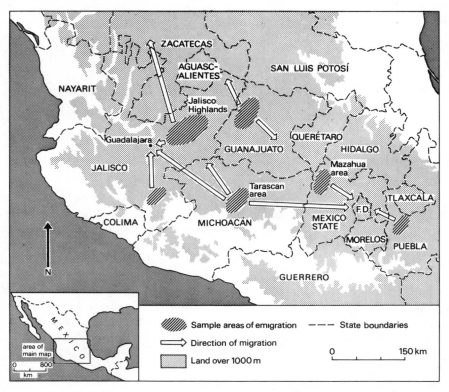

Figure 11.1 The central highlands of Mexico.

rural bourgeoisie. Research shows that these local elements are, in effect, grafting themselves onto a national or international system which governs the majority of social and economic mechanisms but which gives a different appearance to each region. (Figure 11.1 shows the central highlands of Mexico.)

THE PROCESS OF CHANGE

Mexico was the first country in Latin America to experience an agrarian reform. After the Revolution and up to the time of the agrarian reform, the urban middle class claim to have given a new lease of life to the peasantry, whose survival both as producers and consumers was essential to them, within the framework of a dependent capitalist mode of production. To this general scheme, must be added the effects of the varied local political situations which so often are the cause of serious regional imbalances. For example, Government decisions to develop the modernization of agriculture in one region or another are not always a function of strictly economic criteria. Moreover, the agrarian reform has not affected all the regions in the same way. While Morelos, Puebla, Hidalgo, and Querétaro have been strongly affected by the agrarian reform, the same is not true of eastern Jalisco or of Guanjuato, for example, where, on the one hand, structural, and on the other, socio-historical factors come into play. None the less, integrated within a dependent capitalist system, the agrarian reform has evolved everywhere with a certain logic. The problem of the *minifundio* and the division of the *ejidos* arose quickly. The tendency to increase the size of holdings through consolidation became marked in most areas in the 1950s. The change in rôle of the *ejido* from its original function is now common.

While an economic viewpoint emphasizes regional inequalities, change tends to homogenize landscapes and rural people's perception of space and even of their country. It enables the town to organize the countryside which nowadays ensures the dependence of rural people everywhere on likeminded development or commercial organizations, whether public or private. These act as a kind of mediator between town and country. Such new organizations as, for example, expanded banking systems throughout the countryside, or the presence of representatives of development agencies such as the Lerma Plan in Jalisco or the development of the CONASUPO system,* invariably coexist with the traditional forms of Local Government in both the municipal sector, such as the official authorities, and on the level of socio-economic relationships within the group. For example, a large number of peasant farmers has greater recourse to local people willing to lend money, often merchants or *compadres*, than to banks where, in many semi-indigenous zones such as the Mazahua indian region in the State of Mexico (Arizpe, 1975), the farmers are unable to produce the necessary guarantees for the bank loan or else, in the region of the war of the *Cristeros*, they feel a profound mistrust that dates from the time of the Revolution (Rivière d'Arc, 1975). Likewise, in *Los Altos de Jalisco* it has been normal for a long time for surplus produce to be sold either to large- or to small-scale merchants, who are often moneylenders, rather than to CONASUPO where, it is said, too many guarantees are demanded. It should be noted that this hold over the countryside, which arose out of the new structures of rural organization, has had less effect in the old country, in central Mexico, than in the pioneer areas of recent colonization, which is logical but not unique to Mexico. These new structures are superimposed upon the local rural economic organization which, although largely dependent on semi-subsistence farming, is relatively organized. However, in many areas of the central plateau, the new organizations have more limited financial resources than those in the areas of commercial farming. For these reasons, the impact of the reorganization is limited. The infrequency of attempts to establish cooperatives, whether for joint purchase of agricultural products or sale of goods, is noteworthy.

*CONASUPO (Comisión Nacional de Subsistencia Popular) is a State organization responsible for the purchase and storage of grains, and also for the distribution at preferential tariffs of certain foodstuffs.

The spread of agricultural technology, the use of machinery and fertilizers, is the final aspect of economic change in the countryside. From the evidence of published research and personal experiences, it seems that modernization of agriculture is not widespread. Even on commercial farms of 50 ha or more, wooden ploughs pulled by oxen are most frequently used. This is especially true of the *Altos de Jalisco*, where farmers maintain that a tractor would be unsuitable on the stony ground, where bedrock protrudes and the slopes are often very steep. However, this is not the sole explanation, for mechanization is no more advanced in other areas of smallholdings and somewhat larger farms of the traditional area of the central plateau. In addition to the difficulty of farming an area where centuries of intensive cultivation have exhausted the soil, it is impossible for most of the farmers to accumulate even basic equipment. Also, in the *Altos*, Guanajuato, and west central Mexico, generally, the farmers frequently state that an occasional paid labourer is cheaper than mechanization. A farm of 50 ha at certain times of the year may employ fifteen labourers, although workers are becoming increasingly difficult to find. Elsewhere, some innovations were thrust upon the farmers, who, without appropriate preliminary instruction, lacked a full understanding of their use and were unable to understand why their new crops were failures. An *ejidatario* in Zapotlanejo, near Guadalajara, described how he had been ruined through introducing irrigation. He had been advised to try new crop varieties, which had failed completely for two successive years. He also felt he had become a slave to the irrigation for he had to work twice as hard as before in an attempt to get two harvests in a year.

The poor soils, on the other hand, had stimulated a very wide use of fertilizers. Previous research and personal experience in Jalisco, Puebla, State of Mexico, etc., showed that the majority of farmers had used fertilizer at least twice. Fertilizers are least known in the most traditional or remote, often poor, indigenous areas (for example, in the Tarascan indian area—see Pietri, 1975).

The need to increase production and productivity for survival was apparent everywhere as a result of the increased costs of production, in particular the cost of fertilizer. This comes at a time when the local economy is increasingly likely to become part of a badly controlled, ever-expanding system of production with periods of overproduction and times of penury when produce has to be sold cheaply at harvest time when cash is needed and even later to be bought back at a higher price. This would appear to be as true in the traditional just as much as in the modern sectors.

If the production and productivity can only be slightly increased, or not increased at all, other sources of income have to be found. This problem is common over the whole of the central plateau of Mexico but the alternative sources of income vary according to the different socio-cultural and geographical characteristics of the different regions. The alternative is to work in the locality as craftsmen, or for a few months at a time as farm or building labourers, or to become a hawker or artisan, or else to migrate temporarily to the cities, especially to Mexico City or to the areas of commercial agriculture or to the United States.

EMIGRATION

Small-scale farming, rapid population growth (3.5 per cent per year, 1974–75), poor harvests, lack of work in the local towns, and underemployment everywhere eventually make the central plateau an emigration area and at the same time a source of cheap manpower. Thus, the giant metropolis absorbs people continuously in highly centralized heavy industry and in the services that maintain both industry and the urbanization process. Inevitably, the change is not straightforward. Manpower requirements vary according to the stages of growth of the city and according to the degree to which economic change and marginalization go along with integration into urban life. So, various forms of migration develop in response to the tendency to quit agriculture and to become increasingly absorbed into urban employment, in particular temporary migration from the rural areas nearest to the big cities (Arizpe, 1975). It is this, too, to some extent, that ensures the survival of the quasi-rural economy (Rivière d'Arc, 1976).

Contrary to what might be supposed to be the case, it is neither the poorest nor the most marginal communities that are experiencing the greatest emigration, whether temporary or permanent. Rather, it is the areas of traditional agriculture which have the stronger or more long-standing ties with the urban network. The viable threshold for any landholding varies according to region, either in reality or according to the perceptions of local people. Some areas of low population density have annual migration rates that are as high or higher than areas of greater population density (Bataillon, 1975; Unikel et al., 1973).

Mexico experienced rapid and centralized industrialization between 1940 and 1970. This has entailed a polarization of industrial employment, especially in Mexico City, but also in some of the other cities and in certain industrial zones. However, since 1965–70, sociologists from the Colegio de Mexico have observed a concentration of employment opportunities in Mexico City and a decline in its capacity to absorb more people. In spite of this, immigration there has not decreased, except perhaps since 1973, but these data are not yet available. The cumulative effect of this is an increase in non-Government service occupations, of self-employment activities, of artisan work, and of underemployment. Lourdes Arizpe observed that most of the immigrants from the Mazahua indian area spend their months in the town when farmwork in the community is slack, working in the central food market, La Merced, while their wives sold goods in the streets (Arizpe, 1975).

It is hardly possible in this short paper to describe in detail the reasons for migration nor to analyse psycho-sociological or cultural differences which determine regional variations. Everywhere in the central plateau similar economic reasons for migration are put forward. Micro-regional studies confirm the general impression among the rural people that migrants are forced to leave. The poorest people gave the most objective reasons: the minute division of the smallholding or *ejido* among the children or the failure to make credit repayments. However, the poor are not the only ones to leave. In the *Altos de Jalisco*, Michoacán (Pátzcuaro), all the socio-professional classes are affected by migration. The least impoverished feel the work and traditional activities of the small towns to be degrading and long to leave the region. At all levels, the myth of ready money to be found in the cities persists, although the foundation for this belief is not the same everywhere. Mazahua indian farmers attach great importance to education as a means of integration and to a successful life, and they perceive education to be more readily available in the towns. Potential migrants from *Altos de Jalisco*, on the other hand, rarely claim this to be a motive for leaving their community. Short-term migration to the United States has been characteristic of this region, where for more than 50 years, since the *Cristero* wars, the people have felt a profound mistrust of the centralized institutions of the big towns, especially of Mexico City. Emigration to the United States is justified by the possibility of 'getting rich', often with the subsequent intention of returning to Mexico with a little capital.

Seen from another perspective, temporary migration (whether to Mexico, as is general from the State of Mexico, Puebla, Hidalgo, Morelos, etc., or to the United States, from Jalisco, Michoacán, or Guanajuato) has profound effects on the villages or small towns, and often causes violent internal conflicts. In indigenous areas, this change occurs more at an ethno-cultural level, with increased exogenous marriages, giving up traditional language, change in style of dress, a new perception of space, than through more emphasis on social differentiation. In the mestizo or white areas, it is in the relationship between generations, young people and the traditional village authority, that the psychosocial effects of emigration are evident.

These different attitudes to migration have led to the supposition that there are different forms of migration: permanent migration which occurs everywhere, and temporary migration fitted to the demands of local patterns of farming, for a few months at a time allowing frequent visits back to the home village. Temporary migration to the United States now, much more than at the time of official labour contracts to the United States, seems like a real adventure and is valued very highly. These different patterns of migration can be explained by the local historico-geographical situation, but the difference should not allow one to forget that the principal motive for migration is generally economic.

CHANGE, LACK OF CHANGE, AND EMIGRATION

It is not proposed to list the rates of flow of each of these different types of migration: such studies have already been done for the years 1950–1960–1970

Table 11.1

Rate of migration 1960–70, calculated according to the natural growthrate, in three case-study areas		
Tepatitlán (Jalisco)	Pátzcuaro (Michoacán)	Texmelucan (Puebla)
23%	16%	4%
Emigration of young people 14 years old and over (survey conducted 1972–73)		
35%	31%	4%
Maximum family income of the poorest half of the population in pesos (1970) per month		
500	200	200

Source: Bataillon (1975, p. 80).

(Tabah and Cosio, 1970; Unikel et al., 1973). No further quantitative analysis of internal movement in Mexico can be made until 1980.

However, some research data obtained by surveys carried out in 1972 and 1973 (Bataillon, 1975) enable an estimate to be made of the extent of the migratory movement in certain localities in the central plateau. To some extent, these results may be indicative of trends of emigration over a much larger area of the country (see Table 11.1). It can be seen that the migratory rates are very high. However, migration only absorbs a proportion of the natural growth and, overall, the communities of rural central Mexico are not being depopulated in absolute terms, with some few exceptions in the *Altos de Jalisco*, for example. Given the tendency to sympathize with all the facts and mechanisms that explain migration, one might conclude that everyone, or almost everyone, should one day migrate. But that is not the case. Therefore, the analysis of migratory movements should be linked to that of local employment and underemployment, as well as to an analysis of the economic impact of temporary migration on the home community.

René Pietri observed in the Tarascan indian region that, as long as there was the possibility of any supplementary income in the locality, migration was minimal or reduced. Areas that are more integrated into the national economic system still have no greater possibility of alternative sources of income. Talking again of the example of the *Altos de Jalisco*, the town did offer some opportunity of even temporary employment to young people, in petty commerce or in small-scale manufacture of agricultural equipment. Such means of production tend to disappear as they become monopolized by urban industry, more than they would in indigenous rural areas. As a result, emigration from such urban areas is particularly common. Likewise, in areas that participate more in the national economic food production system, local productivity is slightly higher than in the more traditional agricultural zones, costs of agricultural production are high, profits are low, and so migratory flow is particularly strong.

However, it would seem that temporary migration, or even ultimately permanent migration, facilitates at the same time the preservation of a rural economy on which, as has already been observed, the country is reliant. This is achieved by the modest financial contribution of emigrants to rural incomes which allows a low, but constant, level of consumption. On the other hand, where emigration is an individual, or a family, adventure, it encourages people to perceive themselves differently within the local rural group and imbues with a certain fatalism the choice that confronts them: to go or not to go.

The evolution of capitalism in agriculture provokes migration. Migration entails change often in quite specific ways in both a social and a cultural context. This is particularly true of temporary migration.* Permanent migration is particularly conducive to long-term reconcentration of land. Emigration may therefore be at the heart of certain contradictions of the present system, not only in Mexico but in numerous other countries where the problems that occur associated with leaving the country and being absorbed into an urban environment are far from being solved. It would be desirable to slow down the rate of emigration, for obvious reasons, but this would only provide a short-term solution to the problems of the rural people and the least-favoured areas.

In this paper, we have attempted to demonstrate some of the reasons why the central plateau is the region of Mexico most affected by rural emigration; the states of Mexico, Tlaxcala, Hidalgo, Puebla, and Morelos contribute the largest number of migrants, mainly destined for the metropolis. However, at a local scale one is aware that it is not always the most densely populated areas that experience most emigration, and the states on the edge of the central plateau (Guanajuato, Querétaro, Jalisco, and Michoacán) send migrants to a wider variety of places apart from Mexico City, such as Guadalajara, León, and, more recently, Salamanca, where there is a major petroleum refinery, together with the United States. The states in the isthmus of Tehuantepec and the warm tropics (Tabasco, Chiapas, Oaxaca, Guerrero, and Veracruz) experience migration to Mexico City, but also immigration of people bound for the oilfields and the colonization areas. People from the western tropical areas (Colima, Nayarit, and Sinaloa) migrate mainly to the towns on the United States–Mexico border and to the United States, as do people from the northern states of Mexico.

A number of studies have attempted to analyse and document internal migration in Mexico, but have not succeeded in giving a complete picture of the phenomenon since census data do not identify intra-state movements, nor inter-regional migration. Only regional monographs enable one to see the situation at one moment in time, and several studies now under way will enable greater understanding of migration towards the oilfields. This level of analysis does not facilitate the examination of rural emigration in the light of rural development policies of successive Governments. We have already noted that the aims of

---

*In several areas, the returned migrant despises certain tasks and prefers to do nothing rather than lower himself to perform them.

development plans are not always to improve regional economic situations, and political factors frequently assume great importance. The maintenance of the peasantry in their position has enabled them to act as safety valve for the national economy. The widespread persistence of short-term migration to the United States is one means of maintaining the rural population in some zones at least tenuously attached to their rural home. The major infrastructure projects in some regions associated with the provision of irrigation or of land for colonization in the tropical lowlands of, for example, the isthmus of Tehuantepec are generally a response to particular needs and are part of an indirect attempt to check the growth of the regional capital.

It is clear that the phenomenon of rural emigration is not just characteristic of Mexico, nor even of Latin America, but rather of much of the world. It has to be remembered that agrarian reform, whether in Mexico or elsewhere in Latin America, can only temporarily and partially reduce the flow of emigration. Only spontaneous national action, as in Cuba, coupled with a decrease in birthrates can reduce emigration. Without making value judgements concerning Latin American government policies, which would be necessarily superficial, it seems that the first goal must be to offer rural people some choice for their future. This is still sufficiently utopian that it must appear trite.

BIBLIOGRAPHY

Arizpe Schlosser, L., 1975, 'Migration and ethnicity. The Mazahua Indians of Mexico', *PhD Thesis*, University of London.
Bataillon, C., 1975, 'Le départ des migrants mexicains: commentaire à propos de trois études au Mexique central, 1971–74', *Cahiers des Amériques Latines*, **12**, 69–80.
Cahiers des Amériques Latines, **12**, 1975, Numero Special, Migrations au Mexique.
Exter, T. G., 1976, 'Rural community structure and migration: a comparative analysis of Acatic and Acatlan de Juarez in Jalisco, Mexico', *PhD thesis*, Cornell University.
Pietri, A. I., '1975, 'Les mouvements de population au Mexique central, la région de Pátzcuaro', *Thèse de 3ϕ cycle*, University of Paris.
Rivière d'Arc, H., 1976, 'Exploitation familiale, économie rural et migration dans les Altos de Jalisco, Mexique', *Travaux de l'Institut de Géographie de Reims*, **26**, 31–42.
Tabah, L., and Cosio, M. E., 1970, 'Mesure de la migration interne au moyen des recensements, Application au Mexique, *Population*, **2**, 303–46.
Unikel, L., Ruiz, H., and Lazcano, E., 1973, 'Factores de rechazo en la migración rural, 1950–1960', *Demografía y Economía*, 24–58.
Young, C. M., 1976, 'The social setting of migration: factors affecting migration from a Sierra Zapotec village in Oaxaca, Mexico', *PhD thesis*, University of London.

Environment, Society, and Rural Change in Latin America
Edited by D. A. Preston
© 1980 John Wiley & Sons Ltd.

CHAPTER 12

Rural emigration and the future of agriculture in Ecuador

DAVID A. PRESTON

The departure of young people and of entire families from the land has long been a feature of industrialized countries. Although such emigration has sometimes been the result of natural catastrophes, such as the potato blight in Ireland in the 1840s, or of policies directed towards the removal of unwanted people, as in the highland clearances in Scotland in 1782–1820 and 1840–54, it has more often been a steady trickle of people leaving with a more pronounced flow when harvests are bad. In Latin America, the wholesale eviction of tenants and squatters on agricultural land is commonplace and is mentioned elsewhere in this volume, but the general movement of rural people from their homes to live elsewhere is becoming increasingly important. It has already been recognized as a problem in islands of the English-speaking Caribbean, as well as some parts of mainland Latin America.

This essay will attempt to identify some of the characteristics of rural emigration in a single country, Ecuador, and to indicate what impact this emigration has had on farming in a series of sample areas. On this basis, and after comparison with reports of investigations elsewhere in Latin America, some conclusions about the future of rural areas experiencing a high rate of emigration will be drawn and comments made on Government policy towards emigration.

1. GENERAL PATTERNS OF MIGRATION

Latin America as a continent has become increasingly urbanized, and the proportion of the population in urban areas will grow from only 14 per cent in 1920 to a projected 43 per cent in 1980. This has affected countries in different ways and Wilkie considers the case of Argentina, the most urbanized country in the continent. Ecuador is one of the less urbanized countries, with only 35 per cent of the population living in twenty-one towns with more than 20 000 inhabitants, but in 1950 only 21 per cent of the population lived in nine towns with over about 20 000 people. A large proportion of the likely migrants might therefore be expected to be rural people. Ecuador is also distinctive because the area of highest population density, in the Andes, has to the east and west areas with an actively expanding agricultural frontier. Thus, rural emigrants have a choice between urban and rural destinations that are more varied and closer than in the case of many countries. The expansion of petroleum exploitation and exports has stimulated the national economy, as well as causing the construction of roads in parts of the eastern rain forest that were previously not accessible to migrants.

The pattern of rural emigration at a national scale may be simply represented by the identification of the smallest administrative divisions—parishes—where there has been an absolute decline in population between the 1962 and 1974 censuses, although this effectively identifies only those parishes with a particularly sharp decline in population. The distribution of parishes with declining population indicates their concentration in Andean Ecuador—only a few parishes in Manabí on the coast show any population decrease—and Figure 12.1 shows that they are scattered the length of the *Sierra* with the most marked concentrations appearing in Bolívar province south and west of Guaranda and in southern Chimborazo near Alausí. Visits to over thirty such rural parishes in 1975 failed to reveal a very clear regionalization of migration characteristics, and the nature of the mountain environment may mean that adjacent parishes have a completely distinct set of agricultural problems, and within 5 km farmers may suffer crop losses from drought or excessive rainfall.

Studies of rural emigration in different parts of the world have suggested that rural emigration is the result of land scarcity, the failure of harvests as a result of climatic accidents (either frosts, droughts, or floods), and other catastrophic events, such as expulsion by landlords or civil war. Rural emigration is also the result of a series of pressures that prepare families to move and attract them to some other place. The decision to migrate is made up of many different components that affect each family member differently and which make generalizations dangerous. Migration itself must be recognized as a complex phenomenon in which a division between temporary and permanent categories is absurd, since a person may make many journeys for different purposes to the same place lasting different lengths of time and whose permanency may only conclusively be measured on death.

Some simple generalizations may be made that describe general migratory tendencies everywhere and may help focus on the question of the impact of migration on farming. First, most migrants move only short distance and the probability of migration taking place between two locations is inversely proportional to the effective distance between them. This has been indicated in numerous studies and it is elegantly described in relation to Swedish data by Hägerstrand (1957) and the results of other studies are reviewed by Olsson (1965). A second simple theoretical statement is that migrants are attracted to population centres in numbers proportional to the size of that centre. This refers largely to urbanward migration, and no comparable statement can be made

RURAL EMIGRATION AND THE FUTURE OF AGRICULTURE IN ECUADOR 197

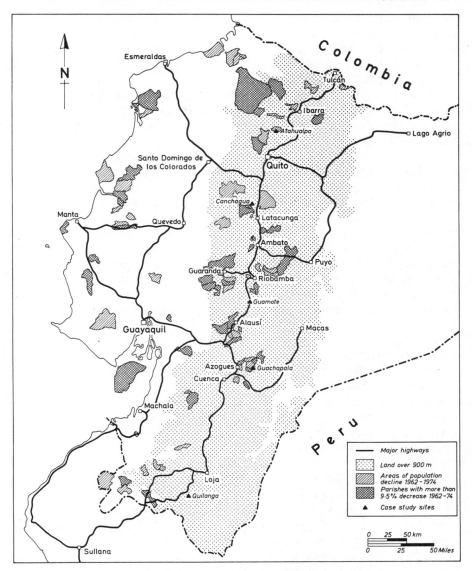

Figure 12.1 Depopulation in Ecuador.

regarding the attraction of different rural areas to migrants. A final statement is that the majority of migrants travel to destinations where they have friends and relatives; thus, present destinations of migrants from a specific source region are in part the reflection of past migrations. Further statements of theoretical importance and which may relate more directly with agricultural chage should focus on who migrates, but considerable variation exists in findings in both Latin America and elsewhere. Migrants seem to include both men and women, but with much local variation, and include both people with no land and with more than enough to support the core of the family.

Remarkably little has been written concerning agricultural change in areas of migration, although the Haneys have documented increased commercialization and associated environmental deterioration in Fomeque, Colombia (Haney and Haney, 1978) and Celestino has reported on agricultural innovation in a rather unusual community in Peru (Celestino, 1972). It is quite possible to construct opposing sets of hypotheses regarding changes that might be expected in areas of out-migration. On the one hand, migrants would deprive the community of the most forward-looking, open, and educated members, and migrants and innovators clearly have many characteristics in common as shown in a review of the body of knowledge of the communication of innovations (see, for example, Rogers and Shoemaker, 1971). As a result, therefore, the residual farming population would comprise those least likely to adopt new methods, and the pace of agricultural change would be slowed. On the other hand, innovation would only be slowed if all or most of the potential innovators left and if most migration was permanent. If, however, much of the migration is short-stay in nature, then the home community accumulates a reservoir of returned migrants who have the attributes of innovators and also a further range of experience, as a result of migration, that may make them more likely to be innovative farmers.

Which of these hypotheses is correct is of considerable importance for national development as well as for the development of rural areas. If areas of small farms, which are characterized by enough emigration to cause the population to decrease, are unlikely to change their agricultural practices in response to the varying social and economic conditions, their output is not likely to increase, and the volume of agricultural produce available for urban consumption will not keep pace with urban growth. An alternative situation could include the growth in size of the middle-sized farms as land released by migrants is acquired, and the gradual penetration into rural areas of capital from outside stimulating commercial farming and banishing self-sufficiency as a goal of local farm production. National policy as conveyed in the 1973–77 Development Plan sees a need to maintain the rural population in place by the provision of 'acceptable levels of productivity which will permit a greater incorporation of the marginalized mass of people which will avoid structural unemployment in the rural areas being transferred to the city' (Ecuador, 1972, p. 7). This is to be achieved as a result of a redistributive land reform, increased land colonization, and by increased efforts to stimulate the production of certain crops and of livestock. It is too early yet to assess the effect of the Development Plan, but agricultural production in the *Sierra* has not changed dramatically in recent years, despite the success of some elements of the Government programme.

2. AGRICULTURAL CHANGE

Agriculture is likely to be affected by emigration in different ways, but to identify which of the many changes in farming practices is linked with migration is more difficult. It is clearly important to discover the nature of agricultural change in areas of emigration in order that it can be seen whether or not the level of living of such rural populations rises, and whether agriculture becomes more or less able to support the farm population. Beyond the actual observable patterns of change in rural communities are the underlying attitudes to agriculture that may influence the approach that individual farmers have to the adoption or rejection of agricultural innovations and to the modification of customary practices. The future of agriculture in the countryside as a whole is, in part, a reflection of the attitudes towards farming held by farmers and other rural people. If, for example, it can be shown that farming is not thought capable of sustaining a family, then it would be unrealistic to expect farmers to invest further in more land, new seeds, or inorganic fertilizers. Thus, what farmers feel about the future of agriculture is at least as important as the availability of new seeds, fertilizers, and equipment in allowing or encouraging rural change and the improvement of rural levels of living.

In order to attempt to discover whether some relationship existed between migration and agricultural change, a series of rural localities that had experienced considerable emigration were studied. The majority of emigration has been from parts of the Andean highlands, and five areas were selected on the basis of preliminary visits in order to represent the range of ecological conditions in which emigration seems to occur. The areas of population decline in the *Sierra* and the localities studied are indicated in Figure 12.1. In each locality, one or a group of civil parishes were used as units of study, and each such parish comprised a central village with 200–300 houses including services such as retail shops, schools, occasionally including a secondary school, medical dispensary, and the offices of local bureaucrats. The majority of the population lived in homes scattered over the countryside, although occasionally in secondary clusters. A random sample of households were interviewed and asked about migration experience and about farming practices and land tenure.

It was found that changes of two types had occurred. In many cases, certain traditional crops were becoming less commonly grown, even some introduced earlier this century, such as rye. While, in other cases, new varieties of crops that were known, but not necessarily widely grown, are now cultivated in conjunction with inorganic fertilizers, insecticides, etc. Farmers indicated that a number of traditional crops, such as quinoa (*chenopodium*) and roots such as oca (*oxalis*), were less commonly grown as dietary preferences changed more towards crops that were consumed by the urban population. Even in indian areas, a decline was noted in the cultivation and use of traditional crops whose consumption by urban people has decreased even though a range of native vegetables, seeds, and tubers is available in major metropolitan supermarkets. It is impossible to link changes in preferences with migration as a whole, since the penetration of different values into rural areas takes place as a result of information diffused by radio programmes, newspapers (for the literate), and by travellers, including many migrants. Migration to urban centres is certainly a potent source of value-forming information, but migration to other rural areas where value systems are less different from those at home is common, and this may diminish the overall effect of migration on community values and preferences.

Changes which involved the adoption of new farming practices were widely known but even so many farmers had not changed their farming perceptibly. Over half of all our informants (54 per cent) had not adopted a single innovation, but in the individual case studies rates of adoption varied between two-thirds in Atahualpa (an area of mestizo smallholders in the *Cordillera* two hours' drive from Quito) and Canchagua (an indian village south of Quito) and only one-quarter in Chantilín (another indian village near Canchagua) and Guachapala (a mestizo village in a mountainous area of southern Ecuador). Although the range of ecological conditions in the case-study areas is considerable, the changes encountered remain remarkably constant, and mechanization is rarely encountered on account of the rugged terrain, and improved varieties of livestock are uncommon in smallholding villages. Sources of advice that can lead to innovation are the provincial offices of the Ministry of Agriculture, and particular extension programmes lead to the promotion of specific crops, not always those about which advice is sought, and many areas are rarely visited by extension agents. Private agencies associated with Church organizations are important change agencies in some places, but not in our sample villages.

The degree of change measured by the number of innovations adopted was compared with three variables for which reliable quantitative data were collected: the area of land farmed, the area owned, and the number of years schooling received by the farmer (Table 12.1). In addition, two variables that

Table 12.1 Characteristics of innovators

	Land owned (ha)		Years of schooling[a]		Age of husband (years)	
	Inno-vator	Non-innovator	Inno-vator	Non-innovator	Inno-vator	Non-innovator
Quilanga	4.21	2.97	2.77	2.93	45.8	39.8
Guachapala	1.59	1.45	2.56	2.10	50.6	53.0
Canchagua	2.04	2.03	2.57	2.17	41.9	38.5
Chantilín	1.20	0.86	4.60	2.63	49.7	55.2
Atahualpa	4.89	1.36	3.14	2.74	45.3	52.6

[a] Of the husband.

Table 12.2 Correlation of innovations with selected variables

	Land owned		Land farmed		Schooling	
	τ	a	τ	a	τ	a
Whole study	0.34	0.001	0.36	0.001	0.13	0.002
Quilanga	0.10	0.2	0.11	0.13	0.001	0.5
Guachapala	0.05	0.2	0.17	0.01	0.13	0.04
Atahualpa	0.38	0.001	0.41	0.001	0.03	0.4

could relate migration in some way to agricultural innovation were used: the previous migration experience of the farmer and whether or not he has considered leaving the area. The analysis, reported elsewhere (Preston, 1977), demonstrates that, while there is overall a highly significant correlation between the area of land farmed and owned and agricultural innovation, and a less significant correlation with years of schooling, when one repeats the analysis at a case-study level, the levels of statistical significance of the correlations are lower (Tables 12.1 and 12.2). A tabulation of innovations with migration intention and migration experience similarly shows a slight degree of significance for the relationship between migration intention, but no significance at a case-study level.

This statistical evidence corresponds well with field experience beyond formal interviews, and we found very few cases where farmers attributed the adoption of specific practices to migration experience and, as we shall consider elsewhere, capital saved during work elsewhere was often not invested in new seeds or fertilizers but rather in equally necessary domestic expenses.

If we can conclude from this that migration has not had a marked effect on agricultural change, it is worth remarking why such a relationship might have been expected. Migration should stimulate innovation because the experience of returned migrants both enlarges personal horizons, which thus encourages the acceptance of new ideas, and provides capital and the means to acquire the seeds, equipment, etc., to change their farming practices. On the other hand, it may be argued more strongly that migration will inhibit innovation because migrants share many of the characteristics of the adopters of innovations (Rogers and Shoemaker, 1971, pp. 185–91) and, in particular, are the most cosmopolite of people—that is those whose 'reference groups are more likely to be outside rather than within their social system'. Thus, migration removes from the community many potential innovators. It may be argued that, so long as there are some potential innovators remaining, this will not seriously jeopardize the rate of change (Griffin, 1976). Previous fieldwork in an area of very high emigration in southern Ecuador showed that extension agents felt that innovation diffusion is hampered by the absence of many men, away working elsewhere, because the women have little decision-making power. This may be, equally, a reflection of the male extension agents' inability to communicate with rural women (Preston, 1974).

A further factor that may inhibit change is the importance of sharecropping as a result of emigration. Under the commonest sharecropping system in highland Ecuador, the harvest is shared equally between tenant and landowner, which does not encourage the tenant to invest his money in expensive inputs for which he receives only half of the produce. Further factors which seem to explain the absence of correlation between migration and innovation include the importance of urban destinations, where migrants lose their involvement in farming and learn little of value for farming life, and the added fact that ruralward migration is most frequently to subtropical or tropical areas where commercial farming of tropical crops is common and teaches migrants little that can directly be applied to their land at home.

It may be instructive to consider one important case where migration had resulted in innovation in the predominantly indian area of Canchagua. Here a number of men had recently obtained jobs in Quito, some two hours' bus ride away, but, since a local bus made the journey regularly, they were able to return home at weekends. They had formed a cooperative in their home community, had hired a tractor collectively, bought tree seedlings to improve a hillside too poor to plough, and were using inorganic fertilizers to improve productivity. Their earnings in Quito were naturally channelled to the community since they returned there at weekends when they did work the women had not done during the week. Their collective identification, not only with farming but also with the community, was reinforced by the availability of money which could be used via the cooperatives for both general and personal benefit.

Agricultural change is the result of purposive action by farmers and of the presence of change agents whereby farmers are informed of alternative possibilities and that further advice is available to those experimenting with new crops, livestock, or farming practices. Since change, by its very nature, involves a new activity, there is uncertainty with regard to its outcome, and thus an element of risk. Where the risk is the absence of an expected harvest for domestic consumption or where the harvest must be sold to enable food to be bought, and where there is little surplus expected other than that necessary to satisfy the basic needs of the farm family, then the consequences of crop failure can be disastrous. In practice, many farmers experiment with new seeds, etc., on a small scale, obtaining only a few new potatoes, or a cupful of wheat, and observe the results in a situation where failure would not be disastrous. As larger-scale experiments are made, farmers with enough land to grow more crops than they need for their own subsistence are thus taking less of a risk than a small-scale subsistence farmer, and it is for this reason that innovative farmers are those who have larger-than-average farms.

The areas where emigration is important are those where the majority of the population works small plots of land, even though large estates also occur. The extension services provide information about possible beneficial changes in farming systems but do not offer a comprehensive range of advice, and the visits of field staff are unpredictable, nor are advice centres available in market towns. This therefore makes agricultural change uncertain and the prevailing small farm size exaggerates the risk of innovations to the farmer, and thus will tend to inhibit change. Since Government policy does not discriminate intentionally between areas of high and low risk and high and low rates of emigration, it can be expected that agricultural change will not in future become more likely in areas of emigration.

3. THE FUTURE OF AGRICULTURE

Rural emigration is an universal phenomenon, but in the areas in highland Ecuador to which we are referring there has been little or no counter-balancing inward flow, and thus a net loss of population occurs since the rate of natural increase is less than the outflow of people. By analysing the situation in a series of highland localities where a very high level of emigration exists, some comments may be made on the possible future of farming in these areas. In some parts of the Caribbean, emigration has led to a decline in the area of land farmed, and economic life has become increasingly dependent on remittances from overseas migrants (see, for example, Crane, 1971; Philpott, 1973). While there is little evidence that this is occurring in Ecuador, it is important to try to forecast what the nature of farming will be where emigration has occurred for a long time. If output per person employed increases and the marketable surplus also increases, then the national economy benefits as argued by Gaude (1976), but, if as a consequence of migration there is little increase in the disposable income of rural families, personal benefits from the emigration of others may be limited and further migration may occur. Continued or increasing migration may well be viewed by the Government with disfavour if it involves a transfer of people predominantly to urban areas, where employment prospects and housing are inadequate, and if there is not an increase in the quality of rural life in proportion to the rate of emigration.

In identifying the view of agriculture held by rural people in a series of Ecuadorian villages, it is necessary to recognize its current rôle in the local and domestic economy. Although, in each of the localities, farming was an activity which employed the majority of the male population, many men and women also worked in a variety of other occupations which frequently involved living elsewhere for periods. In a number of areas elsewhere in Latin America where emigration is considerable, domestic non-agricultural employment has long been important. Thus, in the Mazahua area of central Mexico, families had been accustomed to benefit from employment in mining, road construction, petty trade, and the sale of traditional craft goods (Arizpe, 1976). Sollis (1978) has similarly described the importance of the domestic manufacture of sacks in a small town in Santander, Colombia, as a means of support. In the Ecuadorian communities, cloth weaving, rope making, hat making, and carpentry were all important in different localities and at different periods in the last fifty years, but in every case the importance of this form of employment is much diminished as the importance of factory-made goods has increased and as rural areas have accepted the general urban preference for more standardized goods rather than locally distinctive craft items.

In these communities, therefore, farming has provided the food for the family, even a surplus that can be sent to recent migrants who are establishing themselves in the city, but cash income to buy items in local stores such as salt, cooking oil, and bread, as well as clothing, tools, or medicines, has been provided from non-agricultural sources or from short-term migration. Farming has not, therefore, been regarded as an occupation from which a cash income is derived. Even when extra labour was needed for some farm job, it was often not paid for in cash but in reciprocal labour service. Further evidence of the absence of a cash orientation in farming comes from the range of products which farmers reported having sold. Thus, frequently, farmers said they sold 'nothing', meaning that no crop is produced specifically for sale outside the area, but subsequently informants indicated that they did, in fact, sell at different times any of a wide range of goods, ranging from handfuls of eggs to 6 kg or so of bitter oranges (used for medicinal cures).

It does not seem as though the reaction to diminishing land resources as a result of the division of property between heirs and the lack of unused and available land has been to intensify production and change towards more efficient use of water, to double-crop, and also to grow crops such as vegetables for which there is increasing local demand. Rather it has been to seek non-farm employment either locally or elsewhere. This is largely in response to the example of early migrants who paint a glowing picture of earning potential in the cities or in expanding colonization zones where labour is in relatively short supply and where wages are much higher than in the mountains.

In addition to this, there is a gradual permeation of the rural environment by urban ideas and values as a result of information from the media (especially radio), teachers, visiting Government employees, and returned urban migrants. Each source serves to elevate urban values and to deprecate rural living and patterns of consumption associated with it. Furthermore, children are educated predominantly by urban-oriented teachers to accept a set of values that do not reflect the cultural background of their parents. The same teachers, lacking all but a redimentary knowledge of farming or field-based natural science, are incapable of teaching that part of the school curriculum that demonstrates the elements of farming, rural economics, and environmental studies. The increasing proportion of migrants destined for urban areas during the 1970s further increases urban bias. The proportion of migrants in our case-study areas who had previously migrated to urban areas increased from 18 per cent in 1920–50, to

26 per cent in 1963–69, to 52 per cent in 1974–76 (Preston et al., 1979, p. 23).

An unusually successful savings and loan cooperative in one of our case-study sites (Atahualpa, Pichincha Province) provides, through data on loans authorized, an indication of the investment priorities of rural families. The cooperative includes some relatively poor farmers as well as some people with access to considerable land through rental; a proportion of members of the cooperative (about 10 per cent) are migrants now living in Quito, but the pattern of loans still reveals a fairly clear picture of investment preferences. In 1975, 29 per cent of all loans authorized were for commercial ventures (stocking a shop, buying a welding unit for a blacksmith, etc.), 23 per cent were for dwelling construction or improvement (materials and labour for building on another room or cementing a floor or installing water or electricity), and 22 per cent were for crop farming or livestock (to buy seed and fertilizer, to buy cattle to be fattened). Thus, less than a quarter of all loans are connected with agriculture, providing further evidence beyond the views expressed by our informants that there are several urgent classes of needs that rural folk wish to satisfy before they commit themselves to investment in farming. Beyond this relatively low preference for investment in farming, interviews with both estate owners and small-scale farmers revealed a conviction that their vision of the future for their children and often eventually themselves does *not* include farming but rather some urban occupation.

The priorities that rural people have for the investment of any available capital are predominantly directed towards satisfying their immediate needs for a better home physical environment. Thus, specific questions included in the Atahualpa study showed that 40 per cent of informants would spend any savings from migration earnings on clothing, 20 per cent on everyday living expenses, and 32 per cent on either livestock (20 per cent) or the purchase of more farmland. It may be argued that these data reveal not so much a lack of confidence in farming as a need to improve the quality of life for the family *before* investing in farming. Confidence in farming's capacity to support an adequate level of living is undermined, in many areas visited, by the variability of weather which particularly affects arable farmers and has, in a few places, resulted in greater emphasis on livestock rearing. For most farmers with limited land resources, such a change is impossible, for access to pasture is often limited. While, in some countries such as Mexico, crop insurance is possible and even necessary where loans have been made, in Ecuador such insurance is unknown.

The future of agriculture may be seen in two perspectives. The flow of migrants towards other rural areas implies a faith in faming–but somewhere else—and we have elsewhere demonstrated that migration from rural areas is approximately equal to that from urban areas at a provincial and national level (Preston, 1977). At the level of case studies, although the flow of migrants into the localities was limited, the flow of migrants during the past three decades has been predominantly to rural areas (Preston, 1978) even though urban destinations have recently attracted more migrants. There is, therefore, a substantial body of people who retain a commitment to farming. At a different level, the rôle of agriculture in the livelihood of the population in areas of emigration is probably diminishing as migration earnings become more important and as traditional non-farming employment declines and encourages migration. In 1978, in Quilanga, men were preferring to work as labourers for PREDESUR rather than to plant basic food crops.

3. CONCLUSIONS

Emigration is associated with the perception of agriculture on small farms as offering few chances of adequate financial gain to offset the decline in alternative local non-agricultural sources of income. It is also associated with the rising levels of expected quality of living as a result of the permeation of rural communities by urban values that can only be satisfied by the purchase of more goods. There is thus a need for more money than previously.

We have shown that there is little evidence, in a series of case studies, that emigration is associated with agricultural innovation except in special circumstances, and there is considerable subjective evidence that many farmers are not keen for their children to work in agriculture and are themselves sceptical about their capability to improve their lot as farmers. These elements of peoples' attitudes to farming exist in a physical environment in which a number of alternative farming practices exist which could diversify farming and offer the prospect of better yields for only limited extra inputs of labour or capital. In Quilanga, coffee production could be greatly increased by better care of bushes, improved manual harvesting practices, by simple treatment of diseased plants, and by replanting with better-quality plants, all of which has been demonstrated at Vilcabamba, only one day's walk from Quilanga. In Atahualpa, production could be increased by obtaining access to more irrigation water for low-lying fields and by judicious use of fertilizers, insecticides, and fungicides with root crops and the development of the potential of local fruits for which there is some demand in Quito. In all areas, there is an almost complete absence of appreciation of the value of soil conservation practices, of the value of organic fertilizers, and of rotations that ensure the soil is protected during the wet season and that allow quick-growing legumes to be ploughed in as green manure. This ignorance of soil conservation and measures to improve the organic content of the soil is shared by farmers and extension agents alike.

There are doubtless other measures which could increase output from the existing small farms. It is likely that comparison with the output of South Asian hill farms in a similar climatic zone would further emphasize the considerable increase in production that might be possible from these areas. However, much benefit would be derived from an increase in farm size through a redistributive land reform and the more just apportionment of irrigation water, and improvements in farming methods would also increase income.

The areas of emigration are distinctive not only because their actual population is declining but because few benefits are accruing to their populations either in agricultural innovations, as we have examined here, or in land redistribution, as we have analysed elsewhere (Taveras and Preston, 1977). Because the movement of population out of rural areas is not necessarily beneficial, and because in a number of areas in the Caribbean and in Peru (Castillo, 1964) it has seemed detrimental to the functioning of the community, a special effort seems necessary to encourage experiments with different agricultural systems in order to increase output and thus allow farming to become a more valued occupation to rural people and even to decrease the reliance on frail non-agricultural sources of income. Clearly, this should also be associated with the creation of non-farming employment opportunities to allow more young people to remain in their home community.

Government policy as revealed in the last Development Plan and as currently implemented by the Ministry of Agriculture does not identify areas of emigration as being particularly worthy of attention. The emphasis on specific crops contained in the Development Plan (Ecuador, 1972, Part II, Chapter 1) is concerned with increasing production to decrease dependence on imports and, in practice, conflicts with concern expressed in the preamble to the part of the Plan concerned with agriculture, where social injustice (the unequal distribution of land) is identified as a major cause of emigration, and production cooperatives are suggested as a means of improving well-being, creating a demand for industrial goods, and satisfactory framework for resource conservation (Ecuador, 1972, p. 84). Although the deficiencies of the system of agricultural extension work are clearly indicated, it seems as though, in practice, work has been concentrated on specific crops at the expense of integrated programmes of development. Even Atahualpa, which had been the base of a team of agronomists for a number of year, demonstrated few examples of improved farming except in an area where larger holding existed and where wheat had been very successfully promoted.

It is difficult to be optimistic about the future of agriculture in areas of emigration in highland Ecuador. Although there is undoubtedly the potential for considerable agricultural change of benefit to farmers, there is only limited interest in such change and emigration seems likely to continue without helping to create new opportunities for either non-migrants or returned migrants. The lack of dynamism of agriculture in other zones of emigration in western and southern Europe suggests that agriculture has little chance of providing a better life for farmers without the intervention by outside agencies whose development ideology is oriented to the satisfaction of basic needs through the mobilization of the rural population to design their own development strategy.

ACKNOWLEDGMENTS

Research in Ecuador, on which part of this chapter is based, was financed by the U.K. Ministry of Overseas Development. The views expressed in this paper are however personal and not necessarily those of the Ministry. I am grateful to Rosemary Preston for comments which aided the revision of this paper.

BIBLIOGRAPHY

Arizpe, L., 1976, 'Migration and ethnicity: The Mazahua indians of Mexico', *PhD Dissertation*, University of London.
Castillo, H., 1964, *Mito: The Orphan of its Illustrious Children*, Report No. 4, Andean Indian Research and Development Program, Cornell University, Ithaca.
Celestino, O., 1972, *Migración y Cambio Estructural: la Comunidad de Lampian* (Lima: Instituto de Estudios Peruanos).
Crane, J., 1971, *Educated to Emigrate. The Social Organisation of Saba* (Assen: Royal Vangorcum).
Ecuador, 1972, *Plan Integral de Transformación y Desarrollo 1973–77, Resumen General* (Quito: Junta Nacional de Planificación y Coordinación Economica).
Gaude, J., 1976, *Causes and Repercussions of Rural Migration in Developing Countries*, WEP Working Paper, International Labour Office, Geneva.
Griffin, K., 1976, 'On the emigration of the peasantry', *World Development*, **4**, 353–61.
Hägerstrand, T., 1957, 'Migration and areas. Survey of a sample of Swedish migration fields and hypothetical consideration of their genesis'; in Hannerberg, D., et al. (Eds), *Migration in Sweden*, (Lund: Royal University of Lund), pp. 27–158.
Haney, W. G., and Haney, E. B., 1978, 'Social and economic consequences of rural modernisation in a highland region of Colombia', *Human Organization*, **37**, 225–34.
Olsson, G., 1965, *Distance and human interaction* (Philadelphia: Regional Science Research Institute).
Philpott, S., 1973, *West Indian Migration: the Montserrat case* (London: Athlone Press).
Preston, D. A., 1974, *Emigration and Change: Experience in Southern Ecuador*, Working Paper 52, Department of Geography, University of Leeds.
——1977, *Agricultural Change and Emigration in Highland Ecuador*, Working Paper 187, School of Geography, University of Leeds.
——1978, *Rural Emigration and the Destination of Migrants in Highland Ecuador*, Working Paper 224, School of Geography, University of Leeds.
Preston, D. A., Taveras, G. A., and Preston, R. A., 1979, *Rural Emigration and Agricultural Development in Highland Ecuador. Final Report*, Working Paper 238, School of Geography, University of Leeds.
Rogers, E. M., and Shoemaker, F. F., 1971, *Communication of Innovations: a Cross-Cultural Approach* (New York: Free Press).
Sollis, P. J., 1978, 'Migration and employment in a small town in Colombia. The case of San Gil, Santander', *PhD Dissertation*, University of London.
Taveras, G. A., and Preston, D. A., 1977, *Changes in Land Tenure and Land Distribution as a Result of Rural Emigration in Highland Ecuador*, Working Paper 207, School of Geography, University of Leeds.

External Penetration

Environment, Society, and Rural Change in Latin America
Edited by D. A. Preston
© 1980 John Wiley & Sons Ltd.

CHAPTER 13

The new agrarian and agricultural change trends in Latin America*

Ernest Feder

RUN-AWAY AGRICULTURES

The most fundamental change affecting Latin American agriculture which has taken place since the mid-1960s is the relocation of United States agriculture in Latin America. It is paramount. United States capital and technology, instead of being invested or used in the United States to further United States agriculture, are transferred by individual investors, by agri-business firms supported by their many allies, or by input manufacturers or peddlers into Latin American farm enterprises and agriculture-related industries or services in a whole gamut of 'commodity systems' (as agri-business likes to call them), ranging all the way from common staple foods through luxury items and livestock products to the traditional foreign-controlled tropical and subtropical products, and involving all the necessary inputs to keep the systems running smoothly.

The foreign-dominated commodity systems, often pure and simple aggregations of United States commodity systems, into which industrial capital and technology have been sunk, are always export-oriented, even when the commodities produced are meant to substitute imports, as in the case of staple foods. By this, we imply that to the extent that production, processing, and marketing (including exporting) are controlled by foreign capitalists, the commodities involved are exported or potentially exportable, regardless of domestic needs, the decision on their actual destination, made at the headquarters of the transnational agri-business firms, depending on the relative profitability of the various markets or on other aims. Once installed in a given system, capital and technology can easily be shifted to other commodity systems, (a) in accordance with relative prices (for example, from a staple food for human consumption to animal feeds, or from one type of vegetable crop to another), without however affecting the export orientation of the systems, so that the various systems tend to

---
* The title of this chapter paraphrases that of Andrew Pearse's 'Agrarian Change Trends in Latin America', *Latin American Research Review*, Summer 1966, in which the author analysed prophetically the importance of social change and technology at a time when it was still realistic to speak about 'traditional' agriculture and 'traditional' *latifundismo*.

merge into an undistinguishable mass of transactions where the product as such loses its meaning, or (*b*) in accordance with political and economic pressures which foreign capitalists wish to bring to bear on local economies.

If United States capital and technology are transferred to Latin America and sunk into production, processing, or marketing of commodities which are also produced in the United States, they 'compete' in the United States or world markets with the commodities produced in the United States by farmers there. This is obviously a competition *sui generis*. There is no competition whatever when United States capital controls a commodity system both in the United States and in Latin America. However, when one group of United States capitalists is engaged in such activities while others are doing business only in the United States, then it is not competition between domestic (United States) and foreign investors, but between various rival groups of United States capitalists operating in different localities but selling most likely in the same markets. Where the real competition lies is with respect to producers and rural or industrial workers engaged in the commodity systems involved, since producers and workers in Latin America operate or work for substantially smaller returns or wages than those in the United States and effectively threaten the incomes or jobs of the latter as a consequence of the relocation of the commodity systems.

The relocation of agricultural commodity system parallels almost exactly the relocation of industries from the industrial into underdeveloped—i. e. low-cost—countries. We are therefore in the presence of a common pattern, characteristic of the newest phase of capitalistic expansion, which has been aptly called the new international division of labour and which is characterized by (*a*) the tapping of a practically inexhaustible, cheap labour force, (*b*) the fragmentation of the productive processes, (*c*) the development of cheap transportation and communication technologies, and (*d*) the high geographic mobility of capital and technology. Agricultural production *per se* cannot be fragmented as readily as industrial production, but the various phases of processing and distribution of food and fibre can be so fragmented. In all other respects agricultural relocation and its impact seem identical to the relocation of industries.

AGRI-BUSINESS' SEARCH FOR SUPERPROFITS AND CONTROL

The relocation of United States agriculture obeys a number of closely interrelated economic and political imperatives. As investment opportunities declined on a world-wide scale during the 1960s in the non-agricultural sectors, United States capital searched for alternative opportunities and found them in the previously neglected Third World agricultures. If commodity systems could be controlled in their entirety, from production through marketing and exporting, large profits, even superprofits could be easily obtained. What made foreign ventures most attractive for United States capital, and provided a near-iron-clad guarantee for large profits, was that production and processing facilities as well as improved infra- and suprastructures for these ventures were financed by enormous investment and 'technical assistance' projects of the World Bank and other public agencies, of the 'philanthropic' Foundations (Rockefeller, Ford, Kellogg), and of the United Nations agencies infiltrated by agri-business. They present a pure and simple gift for agri-business firms. Perhaps profits are not as large as in mining or industry because agricultural output is less predictable and prices and supplies are somewhat more difficult to manipulate, but they are nearly as large, so that overseas investments can often be recuperated within one to three years. Besides, United States agri-business concerns had increased their activities within the United States remarkably since the mid-1950s and had accumulated not only large capital reserves for investment elsewhere but also considerable experience in controlling and manipulating United States production and distribution and in manipulating United States farm producers themselves, as for example through the production contract system. They used these reserves for overseas ventures and their experience as well, with the arrogant argument and conviction that they were uniquely qualified to 'develop' the retarded agricultures of the hemisphere. Obviously, the consequences of modernization and mechanization for the agrarian structure in the United States—concentration of ownership and production or processing, the elimination of 'small inefficient' producers and the overall reduction of the labour force—are never mentioned when agri-business tries to convince 'Underdeveloped Governments' of the advantages of modernization agri-business-style, not so much because agri-business wants to hide it from their new customers, but because they see in it one of the miracles which modernization has achieved in the United States, to which they have contributed and which they cherish.

The economic 'push' was accompanied by an equally or even more important political goal: namely to bring 'stability' to the potentially turbulent Latin American agricultures and undermine the possibilities of widespread agrarian reforms, particularly a socialist reform like Cuba's. This has never been publicly acknowledged, but it is hardly necessary: economic control over large sections of Latin American agricultures unavoidably involves their political domination, as foreign investors now step into the shoes of the traditional landed élite at all domestic levels. The success of this strategy has been miraculous. Land reform in Latin America, although more urgently needed than ever before, is now a totally dead issue. In this context, it is noteworthy that the very mentality and performance of the traditional *latifundistas* helped pave the way for the foreign takeover of Latin American agricultures. Their lack of interest in agriculture in general and their failure to invest in their landholdings and help increase output, productivity or employment proved to be a veritable invitation to foreign capitalists to step into their place and 'run the show'. Thus, local land monopolists have betrayed their people and peasants more than once: first, by their failure to use their resources adequately; secondly, by selling out to the foreigners. That the foreigners now betray them again in other ways and still more forcefully has to be added to the original sins of the *latifundistas*.

Overseas agricultural ventures are attracted to Latin America because of extremely low costs all around. They are undeniably the major economic incentive. They include, for example, extremely low wages always inadequate to provide a subsistence income, long working hours for rural and industrial workers including unpaid overtime and work on holidays, the extensive use of women and child labour working at still lower wages or gratis—all in total disregard of national labour legislation and all possible because of an inexhaustible supply of unemployed labour; low land values or rentals; low construction costs; low transportation rates; and the certainty of significant public subsidies, such as price and marketing controls or infrastructural improvements usually financed by external borrowings or free gifts and all the concomitant free facilities to repatriate (super)profits; and United States Government backing for investment ventures gone sour. However, it would be a mistake to look at the costs only and not also at the institutional incentives. Ray Goldberg, the Harvard University guru of United States agri-business abroad, says as follows when referring to the expansion of United States agri-business into the 'farming segment' (i.e. production) in the United States:

> ... the size and public nature of the corporate farmers has resulted in some significant diseconomies. These included the problems of corporate planning and control systems, *vulnerability to union organizing activities* among field and packing-house workers, and susceptibility to *retaliatory actions by independent middlemen* striving to maintain their role in the commodity system ... Both the Justice Department and the Federal Trade Commission are investigating the activities of large corporate farming organizations for possible *violations of the antitrust statutes*. (Emphasis supplied.) (Goldberg, 1974, p. 56)

Here lies a clear hint why United States giant agri-business turns to overseas ventures. In Latin America, there is no need to be afraid of union activities in the field or in processing plants, nor of the action of independent handlers, nor of investigations by Government agencies since they simply do not occur or remain inconsequential. Actually, there are additional facts: rising control in the United States over the indiscriminate use of chemicals, mounting public opposition against contamination, and stricter control over the quality of merchandise sold. It has been argued that the industrial countries shift the production of certain crops and livestock to Latin America as a systematic device to let the underdeveloped populations bear the brunt of contamination. This does not put the issue into the proper perspective. The inveterate opposition of private entrepreneurs to any and all kinds of public controls or competition makes them seek out locations where they can manage their business activities unhindered by such impediments, even if necessary in total disregard of the public interest. The larger the monopoly power of the firms, the greater their contempt of the public interest. In the industrial countries, social progress can thus be cancelled out by the greedy and antisocial search for profits and power of giant corporations, although this is considerably less dramatic than the rapid plundering of Latin America's physical and human resources in agriculture and related sectors, accompanied by the pollution of city and country.

AGRI-BUSINESS—'PARTNERS' OF LATIN AMERICAN AGRICULTURES?

Why do Latin American Governments allow United States and other foreign capitalists to take over crucial sectors of their agricultures under (for them) highly unfavourable, practically catastrophic conditions about which we are going to speak later in more detail?

How do United States (and other) capitalists cajole, cheat or force upon the underdeveloped countries deals which are patently adverse to their apparent global interests?

Who benefits from agri-business in the Latin American economies?

We have already made reference to the poor performance of the *latifundio* sector which, since it controls the majority of the resources and inputs, effectively determines the performance of agriculture as a whole. Agriculture had *grosso modo* a two-tier structure representing an internal division of labour: the *latifundios* including the plantations produced crops or livestock predominantly for export; the peasant sector produced food for the domestic markets. Even when the *latifundio* sector produced partly for a domestic industry, as in the case of sugar, cotton or meat, the related industrial or service sector was likely to be dependent on foreign capital, technology, and markets, if it was not outright owned by foreign firms.

Under existing conditions and with the existing domestic and foreign demand for latifundio-produced goods, it was unneccessary to utilize all the enormous land resources available to the land monopolists. Land could be utilized extensively even for cropping or not utilized at all and still provide very substantial incomes to the *latifundistas*. Private and public investments in agricultures were notoriously insignificant. By the time domestic demand for food and fibre and for rural jobs grew rapidly due to population growth, the *latifundio* sector was structurally incapable and the owners unwilling to shift to land uses more in conformity with the (less renumerative) domestic requirements. Besides, the majority of the land monopolists had only a remote interest as absentee landlords in modernizing their practices, having incomes from other economic activities. To shift to land uses more in response to the new domestic demand for food and jobs would have meant, under the existing conditions, a rupture with the related agricultural and non-agricultural sectors, the banking system, the processors and the exporters or importers—in other words, a break with foreign investors and traders—and a profound, almost revolutionary, transformation in farm management and labour employment practices. It would have meant more intensive cropping, hence more investments and closer supervision on the part of the reluctant *latifundistas* who traditionally preferred to reinvest their agricultural earnings in non-agricultural sectors at home or in industrial countries and in luxurious living; or greater investments on the part of Governments notoriously incapable of raising sufficient taxes for any but the most urgent needs or military equipment. And last but not least, there was then the near-certainty that such a transformation would not be economically attractive for the land monopolists. It is therefore plausible to attribute to the historical dependence of the *latifundio* sector and related industries and services on the industrial countries a great deal of weight as a historical cause for subsequent trends. It no doubt explains the chronic dependence of practically all nations on imports (or 'gifts') of staple foods for domestic consumption to avoid outright famines.

The poor performance of agriculture allowed foreign investors to put forward a reasonable-sounding argument. If the *latifundio* sector would 'modernize' and increase the productivity of the land, food imports could be eliminated, and the production of all kinds of export crops could be encouraged and increase foreign exchange earnings—not a negligible point in view of the exorbitant rise of foreign indebtedness which incidentally coincided in time with the assault of agri-business on Latin American agriculture. From the point of view of the industrial countries, it was anyway preferable to produce the staple foods locally and control their distribution locally rather than to ship them there. Agri-business could even appear as a great benefactor for the suffering masses and for Governments earger to supplement their foreign exchange holdings. No sooner said than done—although it took a great deal to convince traditional *latifundistas* of the advantages of modern farming. (At a later stage, when the foundation of modernization was set, agri-business took over modernization much more directly through acquisition of land and investments in processing and marketing facilities.) To make modernization economically attractive for estate owners and Governments, the convincing was achieved at an enormous expense, particularly for Local Governments, in the form of innumerable direct and indirect subsidies, including the development of the agricultural infrastructure through internationally financed Development Projects from which Governments hoped to recuperate the expenses through export earnings, although it should have been clear that interest payments and amortization of foreign development loans were bound to commit eventual future new foreign exchange earnings from exports well ahead of time, and that the repatriation of profits of private agri-business firms meant another heavy drain on the currency, which shows how illusory the advantages promised by agri-business are. But how is one to reject apparantly generous World Bank, IDB, or Chase Manhattan loans when these agencies are busily engaged in assisting poor countries to 'develop' their economies and simultaneously pave the way for larger agri-business overseas ventures, and when these countries are dependent on these agencies to continue business as usual? Moreover, the United States Government itself was willing to subsidize such an enormous venture which promised to increase future profit repatriations from the new capital and technology transfers. The enormous resources poured initially into the *latifundio* sector achieved their objective to a large extent, although it must be stressed that once the land monopolists had mechanized and modernized, they became more than ever dependent on foreign capital and technology as well as on the erratic commodity markets. Another important part of the persuasion process was the promise that the agrarian structure and particularly labour relations in agriculture would not be altered to the detriment of the large owners, and in fact that modernization could serve to simplify, at least in part, the modernizing *latifundistas*' labour management problems.

Thus, the scarcity of food for domestic consumption and the Governments' increasingly precarious external debt situation, to which 'development assistance' by bilateral or multilateral, private or public agencies contributed systematically, made Latin American economies easy victims for selling out or give away their agricultural resources to United States agri-business. It is not that Latin American political and economic leaders are oblivious to the adverse long-run effects of foreign investments under the conditions under which they are made. They know that for each dollar invested, two or three or four are repatriated back to the United States as profit remittances, and that the difference is shouldered by Latin America's proletariat and costed by its physical resources in an enormous process of plundering; that the argument that a new overseas venture by a giant United States food firm or input peddler 'will save or bring valuable new foreign exchange earnings' is a hoax. They are obliged to accept the deals as a matter of short-run survival. Each new investment postpones the final day of reckoning and the real political victims may be in the next, not their own generation.

Besides, a handful of local estate owners, businessmen, industrialists, dealers, and bureaucrats with great political influence are likely to enrich themselves as 'local partners' of these one-sided deals and to put pressure on their Government to sponsor these transactions. These local partners at the various levels of the agri-business-dominated sectors benefit or hope to benefit substantially from the influx of giant capital and the transfers of technology in various ways which we need not spell out in detail here. Their function is to assist in the smooth operation of the foreign investors' ventures whose management is entirely in the hands of the latter so that the former are 'partners' only in agri-business language. In return for the economic benefits, their rôle, if not obligation, lies in protecting and furthering at the domestic level the status and interests of the foreign capitalists and technology transferrers. This reaches all the way from preventing rises in wages or other cost rates through the prevention of union activities, through the adoption of measures to improve the local investment climate, 'influence peddling' at the Government and administrative levels and image building, to safeguarding the repatriation of (super)profits. Here they are, of course, not newcomers, but they are now strongly backed by the enormous resources of giant agri-business firms and their many allies, and therefore doubly effective. Thus, the local partners, some long associated with foreign capital, provide the 'logistical support' for the activities of the transnational agri-business corporations.

As an example of how the 'partnership' between agri-business and Latin American agriculture actually operates, nothing shows more plastically the complete dependence of Latin American agricultures on the industrial nations than the monopolization of agricultural research of the latter and the systematic strangulation of independent local research in Latin America and other Third World regions. Lester Brown, the first apostle of United States agribusiness abroad, stated as follows in his primitive *Seeds of Change*:

> A predominant share of the World's pool of trained agricultural talent is found within the United States. American agribusiness firms and research stations are by far the most active and innovative. The future course of the agricultural revolution will therefore depend heavily on the ability and willingness of Americans to continue to contribute. (Brown, 1970, p. 72)

This exaggerated statement throws a very misleading light on the issue. A large share of agricultural research in the United States is now focused on crops and livestock grown in Latin America or other underdeveloped regions, under tropical and subtropical conditions for domestic consumption and for export and handled by giant United States agri-business firms. If the research is not carried out in the United States itself, it is undertaken in Latin America by United States Government- or Foundation-sponsored institutions, such as the Rockefeller–Ford–CIMMYT conglomerate in Mexico, or by 'international' research stations sponsored by the Consultative Group on International Agricultural Research (CGIAR), set up by the World Bank, FAO, and UNDP, to which a whole set of well heeled organizations, including the previously named Foundations, and industrial countries contribute funds—an iron-clad guarantee that their work benefits transnational agri-business firms since it is focused on commodities in which they have a vested interest. This strategy obliges Latin American agricultures to look towards United States or United States-sponsored institutions in Latin America for results, but only for the limited range of problems of interest to agri-business. It is logical to argue that this research is in and by itself in support of input sales (new seeds, new breeds, new semen, new equipment, etc.) from the United States to national or foreign producers and processors in Latin America, input sales which are handled by transnational agribusiness (input) concerns, and to control also in this fashion production and distribution in Latin America of commodities handled by United States firms. This entire process does not lack a certain element of absurdity when one recalls that crops and livestock must be adapted to the ecology of specific regions and even sometimes micro-regions where they are to be produced. Much of it ought to be repeated, although usually it is not, so that the flow of repatriatable profits is not unduly hindered by lengthy experiments, regardless of consequences at the local level. All this engenders an enormous waste of resources costed by Latin American agricultures which can be justified only on the grounds that it pushes input sales and achieves control over overseas production and distribution, although it is undertaken in the name of 'agricultural development' in Latin America. The result of this new 'division of research labour' is, on the contrary, effectively to undermine local efforts and development.

THE QUANTUM OF AGRI-BUSINESS IN LATIN AMERICA

It would be useful for our discussion if we could quantify in some detail the transfers of capital and technology from the United States in order to show at least an aggregate figure for the truly gigantic process of the takeover of Latin American agricultures by United States agri-business. This is not yet possible, although there are some statistical bits and pieces which substantiate the claim. The difficulty lies first of all in that these transfers take place at various levels: in the farming sector itself; in processing, i.e. in agriculture-related industries; in distribution and exporting; and in the services related to all of them. Normally, there are no statistics available for foreign investments in farmland and only scattered evidence for the other levels. Another difficulty lies in the widespread use of 'strawmen' to hide real foreign ownership, which is common at all levels. Even an exhaustive study would be unable to throw but a very partial light on the full extent of these transfers.

At the farm levels, the transfers of capital and technology are focused on the best land available in Latin America, in terms of soil quality, climate, availability of irrigation facilities, transportation, and location in general, and on the big farm holdings—which in most cases turn out to be the same thing. This endows the transfers in and by itself with a special sense of importance. At the industrial level, the preference is to acquire already existing firms, unless entirely new facilities are required, while at the service level new facilities are generally needed.

If we keep in mind that foreign capital (and to some extent technology) transfers always imply the marshalling of at least an equal amount of local 'counterpart funds' for all ventures financed by international agencies or private enterprise, and that foreign ventures are demanding in the aggregate a relatively heavy dose of foreign fixed and operating capital, then it follows that the agricultural commodity systems favoured by foreign investors must absorb a disproportionate quantity of (the best) local resources, in relation to the total resources at the disposal of local agriculture. One might say that the agricultural wheels turn entirely around the agri-business-dominated commodity systems and the remainder of agriculture is orphaned. In other words, there simply are no resources left for the peasant sector.

Once the avalanche of foreign capital and technology has been initiated, there is a certainty that more and more 'modern', 'dynamic' commodity systems will be controlled by foreigners so that, in typical avalanche fashion, after a period of time a country's best, most important commodity systems are foreign-controlled. This implies that *a high percentage of total agricultural output and services*—from the farm enterprises through the local supermarkets or export facilities—principally from a relatively small number of large producers, processors and handlers, but involving a relatively large area of farmland, lies outside the effective control of the local economy and its people. This expansion process is necessary for the safe, continued, and increasing repatriation of profits by each individual investor and for the aggregate of foreign capitalists. But the domination of the best commodity systems by foreigners is bound to affect the status and performance of the remainder of agriculture too. The traditional old-fashioned *latifundio* owners who do not wish to go the modernizing way will lose some or all of their economic and political power which served them in the past to obtain the lion's share of the agricultural inputs, and if they discover that income and prestige is no longer attached to the ownership of land, some of them will be inclined or forced to sell out to the modernizers—a peculiar sort of historic retribution which would of course be considerably more attractive if the peasants were to be the beneficiaries. The point to be stressed here is that the effective control of foreign agri-business puts a straitjacket on Local Governments' ability to fashion agrarian and agricultural plans, programmes and projects. Without exaggeration, from now on, agricultural activities in Latin America are managed from the United States headquarters of agri-business corporations and agrarian changes are part of the agri-business corporations' plans only to the extent that they strengthen the land monopolists and weaken the peasant sector. Although this is visible in some countries more than in other, all are now undergoing these 'revolutionary' changes.

The extent of foreign, mainly United States, investment in farmland itself is not a matter of conjecture, except with respect to the exact volume involved. Until a few years ago, land acquisitions by foreign capitalists were relatively modest and some of the giant food concerns, like United Fruit, allegedly sold out their land interests because of the threat of agrarian reforms and adopted other effective and profitable methods to control production and distribution. Now there has been a complete reversal. The amount of land actually purchased by or for United States investors may be estimated to be enormous. In Brazil alone, in 1964–68, according to an unpublished Parliamentary Enquiry in Brazil, over 32 million hectares were acquired by United States investors, over 10 per cent of all farmland. Undoubtedly, additional purchases have since been made. In other countries, reports of large farm sales are common, though in many cases the land is not purchased only for agricultural purposes but to acquire rights to suspected minerals or oil. Leases and concessions provide practically the same long-run control over land and people, and they are no doubt quantitatively even more substantial. In the case of concessions, the main purpose is not to use the land for agriculture, but since they permit control over the access to the land, foreign firms can effectively determine the use of the land and who is going to live on it. Foreign control is also obtained through the so-called assistance extended by bilateral and multilateral development agencies, either through their projects for infrastructural improvements or through projects to foster the production and marketing of specific crops and livestock. For example, if the World Bank authorizes a loan for an irrigation project, this implies not only the use of the land for some commercial crop production, but usually the project is programmed—and controlled—from the very start for the production of itemized commodities normally handled by foreign concerns. Finally, agri-business controls land, land uses, and production or distribution through the production contract system or the purchase of crops in advance of harvest, even without acquiring the land through purchase or rental. Agri-business leaders and apostles like to present these arrangements as a sort of cooperation or coordination between producers and agri-business. But it is no more a coordination device than the arrangement existing between slaves and slave owners under the conditions under which it actually operates, and is in reality an effective new type of vertical integration. Investments in agriculture-related industries and services, whose aggregate and yearly volume is difficult to measure because available statistics are not sufficiently detailed, are made in processing, collection or packing plants (slaughterhouses, canning operations, freezing plants, cold storage, warehouses), input manufacturers or assembly plants (machinery and equipment, fertilizers, chemicals, feed and seed plants, etc.) and a whole gamut of related services (law firms, management consulting, public relations and advertizing, sales outlets) which altogether can be estimated to absorb yearly very high expenditures. The expenditures involve both fixed and operating capital funds. The latter are practically never accounted for. In many cases, they are credits advanced by United States private banks. Quantitatively, they are bound to exceed the fixed capital investments by a substantial margin. Both have the same function: to secure and to maintain ownership or control, or both, over agri-business ventures overseas. The same is true for technology transfers. The influx of capital plus the value of technology transfers at all levels is therefore bound to be substantial each year.

An excellent indicator of the magnitude of these transfers are the loans of the international lending agencies operating in Latin America, the World Bank and the Interamerican Development Bank, since they are practically all in support of transnational agri-business concerns. For example, in 1973/75 Latin America received US $683 million from the World Bank alone, both for agriculture and agricultural industries, but of this total Mexico alone received about US $484 million, i.e. nearly 71 per cent, although the loans were distributed among thirteen countries. The Interamerican Development Bank, in turn, which had lent $2.4 million for agriculture ('directly productive') or 24 per cent of its total lending from 1961 to 1976, authorized in 1976 loans amounting to $428 million or 28 per cent of all loans, though according to my calculations based on a country-by-country listing in the Bank's Annual Report of 1976, the Bank lent $574 million for projects directly or indirectly focused on production (not including projects for general pre-investment studies and planning, for support of exports, rural water, electrification, health, roads or housing, some of which are wholly, others partially devoted to agricultural development). As in the case of the World Bank, the distribution of loans is unequal, since six countries (Argentina, Brazil, Colombia, Honduras, Mexico, and Peru) alone received about $432 million, with Argentina heading the list with $111 million.

Hence, we are not far off the mark if we estimate the annual capital and technology transfers at between one and two billion dollars during the last ten to twelve years, the lower figure being applicable to the first five to six years, the higher figure to the last years. This includes investments from private and public sources and all sales of technology (in the broadest sense). Perhaps the figure is conservative. If these investments return 30 per cent as profits—a conservative figure—we can obtain a sense of the plundering of agricultural resources in Latin America by agri-business. To this, must be added numerous long amortized investments, such as the foreign-owned plantations, for which the returns are pure profits.

THE LONGER-RUN ECONOMIC PERSPECTIVE

If one assumes that the process which I have described will continue at an accelerated rate, which is realistic, it will have the most far-reaching consequences.

The first question is whether the concentration of ownership and production at the farm and industrial or service level, the almost complete mechanization of the productive processes, the depopulation of the countryside, all accompanying an increasing productivity of land, men and beasts, characteristic of the agricultural and agrarian structure dominated by agri-business in the United States, will all be transferred together with United States capital and technology to Latin America. Our answer is a practically unqualified 'yes'. There is no reason to suspect that in underdeveloped dependent agricultures the process of increasing monopolization would not operate exactly like in the industrial countries. There are, in fact, reasons why it will operate even faster. United States agri-business encounters in the hemisphere an already exceedingly high level of concentration which has tended to increase at a steady rate even prior to the entrance of United

States agri-business. Trends towards greater monopolization of land are now brutally reinforced by the entrance of United States agri-business: on one side, by the systematic discrimination of United States investors against the small (peasant) holdings; on the other, by the indiscriminate and uncontrolled use of labour-saving technologies; and, thirdly, by the immense resources of United States capitalists which, once they are installed, allow them to monopolize land to a still larger extent than the traditional and modernizing landowners or to help satisfy the land hunger of the local monopolists, thereby systemetically undermining the means of livelihood of the peasants.

The same occurs in agriculture-related industries and services. The two trends—rising monopolization at the farm and industrial or service levels—reinforce each other mutually. The incessant pressure to use ever more sophisticated technologies, including labour-saving machinery and equipment, and modern farm or plant management methods, all well adapted to needs of large-scale modernizing *latifundios*, greatly adds to the dilemma. The necessity to use the equipment at full capacity under modernized conditions must lead to an expansion of individual modernizing national or foreign-owned farm and non-farm enterprises, exactly as it did in the industrial countries. Technology transfers *per se* are instrumental in transferring the socio-economic structure (production relations) of their place of origin under the conditions under which they are now made. The technocratic madness of this search for automation in agriculture can be illustrated by efforts to introduce electronic equipment in agriculture in India, one of the world's most impoverished countries. In a report of Panel on Agroelectronic Instruments we read as follows:

> The increasing stress on mechanization of agricultural operation to increase the productivity requires the use of instrumentation. Electronic instrumentation is an effective tool for immediate improvements in many operations. There are many areas where electronic instrumentation can be introduced.

The report then enumerates a set of activities for the farm and product handling levels which can easily be automated electronically. This is not an isolated case of madness; it symbolizes a system gone mad.

What does it all mean when we look at it from the point of view of the smallholders and the rural or industrial wage workers? Two examples seem to be pertinent. In West Germany, where agri-business United States-style is only in an early stage of development, between 1950 and 1972/73, the size of the family labour force decreased by roughly two-thirds, and the wage labour force decreased as a result of mechanization by over five-sixths, whereas the number of farm enterprises over 0.5 hectares declined by about one-half, from roughly 2 million to about 1.1 million farms, and of farms over 2 hectares from 1.3 to 0.9 million; in 1974 it had further decreased to 0.8 million and the elimination of the small units continues. In absolute numbers over 3.5 million people left agriculture during the twenty-two-year period. In the United States, during the same period, rural-urban migration amounted to some 13.5 million people. Clearly, the eviction of small producers and wage workers accelerated sharply since 1960. On the other hand, in Mexico where United States agri-business capital and technology are rampant, the rural labour force decreased by 17 per cent between 1960 and 1970, but the number of *ejidatarios* (smallholders) alone decreased by 47 per cent. The mathematically inclined reader can easily predict how many decades it will take for Mexico's most important peasant sector—the *ejidatarios*—to disappear from the agricultural scene if present trends are allowed to continue.

Thus, the handwriting is on the wall. I conclude that the elimination of the peasants and rural wage workers—the logical result of modernization United States-style—is immensely speeded up by run-away agriculture. Although it would be absurd to argue that the millions of peasants and workers will disappear from the agricultural sector in a matter of years, it must be taken for granted that most rural job opportunities are now threatened and will have disappeared in Latin America within two or three decades, and those new unemployed will find no or only insignificant job opportunities in agriculture-related industries and services, given the already existing enormous unemployment in all Latin American economies.

But why must our answer be qualified?

Agri-business comes to Latin America in order to make superprofits. Superprofits result when costs are and can remain super-low. Among other things, costs of production are low when agricultural resources can be exploited mercilessly without any obligation to use them effectively and so as to conserve them. Foreign capitalists do exploit agricultural resources in the hemisphere mercilessly. They are under no obligation to maintain or help maintain drainage systems to prevent salinization of irrigated areas, or maintain irrigation facilities to prevent water wastes, or reforest deforested areas. Modernization of Latin American agricultures is a one-sided affair in which modern technologies are used *selectively* to secure superprofits, but where the technologies to maintain the capital base of the overseas ventures do not enter, as they do in the industrial countries, because this would raise costs of operation. Thus, the foreign capitalists operate exactly like the traditional *latifundistas* except at a much higher level of technification and therefore with considerably more destructive results. The destruction can continue as long as there is still an abundance of land, water and forests to be exploited. If the resources are destroyed in one area, it is simple and cheap to move to another where resources are still intact and where the same process can be repeated. Hence, agricultural output in Latin America may rise but at the visible and increasing cost of future outputs. This is certain to leave Latin America dependent for food supplies on the industrial food-exporting nations for ever.

THE NEW AGRARIAN AND AGRICULTURAL CHANGE TRENDS

Foreign agri-business' control over production, distribution, and destination of the output now introduces an element of instability which even with—or particularly with—a rising initial output will grow by leaps and bounds. Distribution and destination being determined by prices in local, United States or world markets, foreign food monopolists will sell in the most profitable markets, which may leave the domestic markets without adequate staple food supplies, forcing them to import more expensive ones at a later time; it may leave them with unusable supplies of products for which the purchasing power of the domestic market is inadequate in case of a commodity crisis in United States or world markets. (The latter happened recently with respect to cattle and meat for example.) This results in extraordinary resource wastes. If the output is not yet produced, changing relative prices affect land uses and lead to constant shifts from less to more profitable commodities, for example from one staple food to another, from staple foods to animal feeds, or from staple foods to luxury items, or vice versa. Such shifts have more dangerous longer-run implications for the local food supplies and nutrition than the short-run search for the most profitable markets for supplies already produced. We must keep in mind that the takeover by agri-business of all important food and fibre commodity systems abolishes the previously mentioned two-tier production structure with its built-in relative stability of food supplies for domestic consumption, so that the nutritional situation becomes increasingly more problematic, even chaotic, since the satisfaction of local needs has a lower priority for transnational agri-business firms than the need to repatriate superprofits. If underdeveloped countries now import more basic foods in the face of larger and larger transfers of foreign capital and technology, we must seek the cause for these trends in the world-wide activities of the transnational agri-business concerns supported by international development assistance agencies.

The increasing dependency and intertwining of Latin American agricultures on and with world markets must result in wider and wilder price and income variations in all local markets, as economists well know. The price fluctuations in world food and fibre commodity markets have sharply increased, in the last ten years. For some United States authors such as Lester Brown, the fault lies with the Soviet Government, with Governments meddling with prices and supplies. Brown says in a recent pamphlet:

> The Soviet decision to offset crop shortfalls with massive imports rather than via the more traditional method of belt-tightening by consumers is the most destabilizing factor in the world food economy today, one which is enormously costly to consumers everywhere. The instability derives not so much from the scale of Soviet grain imports as from their unpredictable and secretive nature. (Brown, 1975, p. 18)

This is an absurd thesis for a variety of reasons. It puts the capitalist order in an implausibly bad light if the little tail (the Soviet Union) can wag the big dog (the

United States), given that the latter controls practically the entire world wheat market. Many commodity markets are chaotic in which neither the Soviet nor the Chinese have any participation or interest whatever. But Brown is not one to put the finger on the United States food monopolists, as he ought to, since he is the transnational agri-business concerns' staunchest supporter. Actually, the food monopolists can successfully unload the economic burden of unstable markets and food crises on the 'natives' and the largest portion of the costs of adverse market conditions are paid by the small producers (smallholders) whenever they supply part of the output. In the face of all this economic chaos and the existing and expected brusque changes at all levels, Latin American countries are totally helpless.

THE POLITICAL PERSPECTIVE

The foreseeable depopulation of the countryside is certain to have a far-reaching political impact. If Latin American agricultures can be run without peasants and with only a token labour force of wage workers as the number of labour-intensive commodity systems declines with rising automation, the rural population will cease to be a meaningful political element in society. Agri-business-type modernization is a counter-reform strategy to prevent peasants from obtaining land. But it aims much farther. The exodus of the rural proletariat will make any future peasant or workers' organization, movement or protest impossible for lack of people, and there will be no more 'peasant wars' of the kind described by Eric Wolf. Thus, the class conflicts are shifting entirely to the cities whose slums can probably be maintained in check politically and militarily with relative ease.

It now seems that the expansion of capitalism in the underdeveloped agricultures leads to such an immediate array of practically insoluble dilemmas that the industrial capitalists will be no more able to control them than the sorecerer's apprentice the floods of waters he conjured up. The dilemmas in Latin America are already enormous: more agricultural commodities—more hunger; more modernization and technology—more unemployment and poverty; more rural exodus—more urban slums. The end is not in sight. Until now, the response to these dilemmas have been schemes to 'assist the rural poor' whose only possible outcome must be the proliferation of the poor; birth control and sterilization projects which have no bearing whatever on the poverty problem; and other measures directed towards some manifestations, not the roots of the problems. The only real alternative is for the capitalists of the industrial countries, who have always shown a capacity to turn to their advantage the existing economic, social, and political contradictions which their system generates, to turn to a radically different strategy of development which would be of real benefit to the poor. It would have to envisage a large-scale programme of redistribution of income and wealth and protagonize labour-intensive agricultures and rural industries. Such a plan has not come forward. What has come forward is a scheme so mild, inoccuous and full of false pretences that one has the right to question whether the capitalist system will this time be able to master the dilemmas mentioned above. I am referring to the joint proposals of the world Bank and the Institute of Development Studies (Sussex) under the title *Redistribution with Growth*, hailed in some quarters and by the authors themselves as a revolutionary, 'far-out' new development strategy, and whose importance lies in that it provides the theory for new programmes such as the poor assistance schemes of the World Bank started in 1973. Because of their uniqueness, a brief comment at this point may put the future of rural Latin America in the proper perspective. Its centrepiece is the argument that a mild redistribution of income may not be an obstacle to growth (larger GNP). Now income redistribution is impossible without a redistribution of wealth. But the authors are not in favour of the latter. They claim that 'Political resistance to policies of asset redistribution makes this approach unlikely to succeed on any large scale in most countries'. (Chenery et al., 1974, p. 49)

This could be true for dependent capitalist, but not for socialist countries. In the latter the 'political resistance' was overcome. When examining two approaches to asset redistribution: the 'static' redistribution of an existing stock of assets, and the 'dynamic' approach aiming at altering patterns of asset accumulation over time, the authors state with respect to the first that it is 'the most immediate and also most radical approach' and immediately question 'how far . . . such measures [are] likely to be effective in redistributing income assuming that they are politically feasible'.

In other words, the capitalist system is not in favour of such a 'radical' approach. But what about the dynamic approach, the redistribution of new assets?

> . . . in espousing the general principle of redistribution the benefits of growth, an essentially political judgment was made [by the authors] which is thematic [sic] to this volume. This is that intervention which alters the distribution of the *increment to the overall capital stock* and income will arouse less hostility from the rich than transfers which bite into their existing assets and incomes. (Chenery et al., 1974, p. 56)

Unfortunately, this proposal suffers from a serious defect: it is unrealizable. If new land is taken into production, a small portion could perhaps be turned over to many smallholders while the remainder goes to a few land monopolists. (Actually, the authors' conception of land redistribution to the peasants is so nebulous that it remains a far cry from land reform and reinforces our argument that one can forget about the authors' asset redistribution scheme altogether.) But the give-away of a portion, no matter how small, of a new processing plant, a rural highway or an irrigation dam? If redistribution of wealth is unrealizable under the *Redistribution with Growth* strategy, it is bound to throw a serious question on how income redistribution can be achieved. Let us look into this matter, as the authors see it. The ultimate purpose of the new strategy is, as we said earlier, to have growth with a mild redistribution of income. Here are two crucial quotes:

> Our diagnosis of the nature of poverty in developing countries and of the inadequacies of conventional measures of social progress leads us to propose a *fundamental redirection of development strategy*. (Chenery et al., 1974, p. xvii)

What do the authors understand by a fundamental redirection?

> A strategy involving the annual transfer of *some 2 percent* of GNP from the rich to the poor *for one or two decades* would not be accepted readily by the rich. (Emphasis supplied) (Chenery et al., 1974, p. 52)

The authors' horizon is thus very modest to start out with. A 2 per cent transfer annually of GNP may involve the percentage with reference to the total or to the added GNP, always on the assumption of a yearly increase in GNP. As we already know, the authors are in favour of the second approach. If we use the simplified example below (Table 13.1), a 2 per cent transfer of total GNP would be a one-shot redistribution affair, with the share of the poor increasing once and for all from 20 to 22 per cent. If one uses the added GNP as a base, the transfer would result in a gradual increase in the share of the poor, but in smaller and smaller instalments, reflecting smaller percentage increases in GNP (annual increases of $100). Even if the scheme would continue 'for one or two decades' in this example, the overall effect would be on the skinny side. It would not matter whether one adopts case I or case II method. Practically it is no redistribution at all. Hence the authors prefer the case II methods as a simple political 'gimmick' because it gives the illusion of a rising share of the poor in the national income over time. If GNP would increase more normally, say by 5–10 per cent, the effect would of course be proportionally smaller. Once the added share of GNP is divided up among the myriads of poor, the latter would never realize that they are supposed to be better off, even if we forget that the poor multiply faster than the rich. Actually, we need hardly make the effort of providing an example. *Redistribution with Growth* provides us with a forty-year model of income redistribution (p. 219ff) whose effectiveness is reduced not only because the now-living poor will be dead after forty years but also because it shows even more meagre results than our own example: after forty years, the share of the poor of GNP is the same as in year 0. Thus the modest 2 per cent scheme is nothing but a statistical game and the authors give the show away by stating that the rich would 'not accept it readily'—a refined way of saying that the rich would fight it tooth and nail. And why should the rich not fight even such an ultramodest scheme

Table 13.1 World Bank-type annual transfer of 2 per cent of GNP

Period	GNP	Case I				Share of the poor ($m)	(%)	Share of the Rich ($m)	%
		Share of the Poor ($m)	(%)	Share of the rich ($m)	(%)				
I	100	20	20.0	80	80.0	20	20.0	80	80.0
II	200	44	22.0	156	78.0	42	21.0	158	79.0
III	300	66	22.0	234	78.0	64	21.3	246	78.7
IV	400	88	22.0	312	78.0	86	21.5	314	78.5
V	500	110	22.0	390	78.0	108	21.6	392	78.4

Note: Case I: transfer each year of 2 per cent of total GNP. Case II: transfer each year of 2 per cent of added GNP with an ultramodest goal which does not include any asset redistribution worth mentioning so that it could not even be implemented? All their efforts are spent today as yesterday on devising mechanisms to shift an ever-increasing portion of the national income and wealth to themselves. Another weakness of the scheme is that it is tied to an *increasing* GNP. Neither God nor the Devil himself could convince the rich to turn over a portion of their income during a depression. Or are the authors perhaps proposing that if GNP declines the poor should now reimburse the rich at the rate of 2 per cent of the decline?

I have commented on this anything-but-revolutionary new development scheme because it demonstrates that the new change trends in agriculture which I have described: the rapid concentration of landownership and production, the concentration in the agriculture-related industry and service sectors, the depopulation of the countryside, the growing chaos in the local food supply, price and income situations, the growing dependence of Latin America on the United States, all consequences of capitalist expansion implemented by gigantic capital and technology transfers operated by agri-business transnational concerns and their powerful allies—that these trends must lead to ever-greater social, economic and political imbalances if they are allowed to continue. Obviously a redistribution-with-growth-mentality and schemes innoculated with it even hasten these imbalances. They are all the 'more enlightened' capitalists have been able to come up with, and they are regarded as 'far-out' and actively opposed by other leading financiers and industrialists. If we were to ask the (politically naive) question whether the capitalist system can come forward with a more effective scheme, not a programme which 'assists the rural poor' but which will do away with them, provide jobs, food and basic needs and take some away from the rich and give it to the poor, or whether the system has perhaps run out of ideas on how to stem the chaos, we would have to answer 'no' to the first and 'yes' to the second part of the question, and this is after all an optimistic way of looking at it when one disregards the tremendous individual and social costs associated with this process of change.

SUPPORTING LITERATURE

Agribusiness Accountability Project, 1973, *A Summary Report on Major US Corporations Involved in Agribusiness*, Washington, D.C.

Barraclough, S., 1973, *Agrarian Structure in Latin America* (Lexington, mass.: D. C. Heath).

——1977, 'Agricultural development prospects in Latin America', *World Development*, **5**, Nos. 5–7.

Breman, J., 1974, *Patronage and Exploitation* (Berkeley: University of California Press).

Brown, L., 1970, *Seeds of Change* (London: Pall Mall Press).

——1975, 'The politics and responsibility of the North American breadbasket', *Worldwatch Paper* 2 (Washington, D. C.: Worldwatch Institute).

Chenery, H., et al., 1974, *Redistribution with Growth* (Oxford: Oxford University Press).

Electronics, Information and Planning, 1977, New Delhi.

Ernest, D., 1977, 'Technological dependence and development strategies', *Lund Letter on Science and Technology*, No. 3.
Feder, E., (Ed.), 1973, *La Lucha de Classes en el Campo* (Mexico D. F.: Fondo de Cultura Económica).
Feder, E., 1976, 'How agribusiness operates in underdeveloped agricultures: myths and reality', *Development and Change*, 7, 413ff.
—— 1976, 'McNamara's Little Green Revolution, World Bank scheme for self-liquidation of Third World peasantry', *Economic and Political Weekly* (Bombay), 11, No. 14; and *Comercio Exterior* (Maxico), 22, No. 8.
—— 1977, 'Capitalism's last-ditch effort to save underdeveloped agricultures, international agribusiness, the World Bank and the rural poor', *Journal of Contemporary Asia*, 7, No. 1.
—— 1977, *Strawberry Imperialism, An Enquiry into the Mechanisms of Dependency in Mexican Agriculture* (Mexico D. F.: Editorial Campesina) (distributed by America Latina, London).
—— 1977, 'Agribusiness and the elimination of Latin America's rural proletariat', *World Development*, 5, Nos. 5-7.
—— 1977/78, 'Campesinistas y descampesinistas. Tres enfoques divergentes (no incompatibles) sobre la destrucción del campesinado', *Comercio Exterior* (Mexico), 27, No. 12; and 28, No. 1.
Froebel, F., Heinrichs, J., and Kreye, O., 1977, *Die neue internationale Arbeitsteilung* (Hamburg: Rowohlt).
Funk, A., 1977, *Abschied von der Provinz* (Stuttgart: Plakat Bauernverlag).
Goldberg, R. A., 1974, *Agribusiness Management for Developing Countries—Latin America* (New York: Praeger).
Hewitt de, Alcántara, C. 1976, *Modernizing Mexican Agriculture* (Geneva: UNRISD).
Jacoby, E., 1974, 'Structural changes in Third World agricultures as a result of neo-Capitalist developments', *The Developing Economies*, 12, No. 3.
Moore Lappé, F., and Collins, J., 1977, *Food First* (Boston: Houghton Mifflin).
Morrissy, J. D., 1974, *Agricultural Modernization through Production Contracting* (New York: Praeger).
Pearse, A., 1977, Technology and peasant production, *Development and Change* (The Hague), 8, 140ff.
—— 1976, *The Latin American Peasant* (London: Frank Cass).
Rott, R., 1977, *Strukturelle Heterogenität und Modernisierung*, Sociology Department, Freie Universität Berlin (to be published).
Tuomi, H., 1976, 'On food imports and neocolonialism'; in *Political Economy of Food, Proceedings of an International Seminar*, Tampere Peace Research Institute, Research Reports No. 12, Tampere (to be published by Teakfield, London).
Wallensteen, P., 1976, 'Scarce goods as political weapon: the case of food'; in *Political Economy of Food, Proceedings of an International Seminar*, Tampere Peace Research Institute, Research Reports No. 12, Tampere (to be published by Teakfield, London).
Wolf, E., 1969, *Peasant Wars of the Twentieth Century* (London: Faber).

Conclusion

Environment, Society, and Rural Change in Latin America
Edited by D. A. Preston
© 1980 John Wiley & Sons Ltd.

CHAPTER 14

Some concluding comments: directive change and the question of participation

NORMAN LONG

The preceding essays offer a panoramic view of Latin American rural development problems and processes. From this, one can isolate three broad themes that are significant for discussions of participation and development. The first concerns the analysis of the characteristics and outcomes of different types of rural development policy, paying particular attention to situations of major State intervention. As several contributors point out, large-scale rural development programmes, involving for example land reform or colonization schemes, must not be judged successful solely by reference to the establishment of efficient systems of production and administration that maximize economic return and generate high value for investment. Consideration must also be given to the types and levels of material and social benefits enjoyed by the peasant beneficiaries or the settlers themselves. Hence, the evaluation of development programmes is crucially linked to questions of equity and grassroots participation.

A second related theme has been the differential responses to economic change and to programmes of development shown by different segments or classes of the rural population. This issue emphasizes that 'participation' is fundamentally *interactive* and *competitive*. Political struggles, both within the peasantry and between peasants and other social classes, focus on obtaining preferential access to the benefits of development, whether these be productive resources, jobs in the bureaucracy, or political support from regional or central Authorities. Under certain conditions, local groups (e.g. commercial farmers, merchants and traders, or organized peasant groupings) may themselves modify or even undermine the best-laid plans and priorities of Government. The formulation and implementation of development policy is, therefore, an ongoing political process whose characteristics and outcomes may look very different from the perspectives of the various actors or power-groups involved. This heterogeneity and clash of social interest provides an important analytical focus for interpreting many of the development situations described in earlier chapters.

A third major dimension emphasized in the volume is the need, particularly when discussing Latin American examples, to take full account of the historical development and external orientation of Third World economies. Although scholars may disagree as to the timing, extent, and consequences of foreign capitalist penetration, Latin America is notable for its long and varied history of colonial rule and early incorporation into the international economy. At different historical junctures, and in different ways, Latin American societies have had to respond to the needs of the economically more powerful Western nations. One of the latest and perhaps most systematic forms of foreign domination and exploitation is that exercized by multinational companies and international aid agencies in the promotion of the production and processing of agricultural foodstuffs mainly for export. This process is having far-reaching consequences for the rural inhabitants of regions of Latin America, and thus, like early types of intervention, must be considered of crucial significance for structuring the economic and political alternatives available to the different rural classes.

These are complex and interrelated problems for analysis. Clearly it would be somewhat contrived, and probably foolhardy, of me to attempt a synthesis of the varied materials and arguments relating to the above themes that are presented in this collection. Instead, I intend to delineate some of the principal characteristics and consequences of increased intervention and centralization by the State in rural development in Latin America. This process has major implications for understanding the meaning and problems of 'participation', particularly as manifested in State-sponsored development programmes that explicitly adopt a 'directive' as against a 'non-directive' approach to planning and implementation. I also briefly suggest that an actor-oriented perspective on development issues is useful for dealing with problems of differential responses to change and for discussions on participation.

THE MAIN CHARACTERISTICS OF DIRECTIVE CHANGE

Discussions of rural development policy often distinguish broadly between 'improvement' and 'transformation' approaches. Whereas the former aims to encourage agricultural development within existing peasant production systems and introduces relatively few changes in the legal and social structure of rural society, the latter makes a sharp break with existing institutional forms in order to establish new, and generally larger-scale, systems of production and organization, often involving new patterns of land tenure or new types of rural settlement. Examples of the improvement approach are agricultural extension and training programmes organized for peasant farmers, and various 'self-help' projects associated with a Community Development or *animation rurale* strategy (see Long, 1977, pp. 145–158; Coombs and Ahmed, 1974, pp. 66–88). Characteristic of the transformation approach are programmes of land reform, State-organized colonization projects, and population resettlement schemes (see Long, 1977, pp. 158–81; Lehmann, 1974; Chambers, 1969).

Although, in certain respects, the distinction I wish to make between 'non-directive' and 'directive' change bears some similarity to the contrast between improvement and transformation strategies, this distinction (originally suggested by Batten, 1974) places greater emphasis on the organizational and participatory aspects of development projects rather than on their more general economic and structural features. Also, whereas the difference between improvement and transformation stresses the goals or ends towards which rural development is oriented, the contrast between non-directive and directive change emphasizes the administrative methods and organizational means by which policy goals are to be formulated and achieved. Hence, the main criterion for differentiating non-directive from directive change is the degree of grassroots, voluntary participation in the design and implementation of development projects, something of course that the philosophy of Community Development or *animation rurale* also stressed.*

Directive change is characterized by a pattern of centralized control and administration by which the objectives and means of implementation are determined by Government or the sponsoring agency. Whilst this does not exclude the possibility of the beneficiaries or settlers participating in some decisions concerning the day-to-day operation of the production process, all the major decisions and the overall organizational framework remain firmly in the hands of State officials or representatives of the sponsoring body. Thus, the State or agency officials determine the main cropping pattern, the technology used, the management of agricultural inputs and credit, and the arrangements for marketing. In addition, Government statutes or organizational regulations usually define the precise criteria for membership of the project, and the rules for the allocation of land and water, and for the distribution of returns from production.

In contrast, the non-directive approach permits much greater involvement of various interest groups in determining the objectives and nature of the development strategy to be adopted. By emphasizing the rights of the various parties (including the peasant farmers or landless labourers) to formulate and press for the adoption of policies which they see as furthering or protecting their own particular interests, this approach can lead to bargaining and sometimes to open political confrontation between competing or antagonistic groups. It can also result in a lack of consistency in the formulation and implementation of policy; and may have the disadvantage that the less organized, less articulate, and materially more deprived sectors find themselves in an even weaker bargaining position, unless they can develop their own organizational and political skills, than they might otherwise be under a directive strategy. Indeed, it

---

*The following discussion of the main features and shortcomings of the directive approach draws from an earlier paper by myself and David Winder presented at the *Xth Congress of the European Society for Rural Sociology*, Córdoba, Spain, April 1979.

is usually necessary for the State to give some formal recognition, and sometimes direct political support, to the weaker or poorer groups so that they are able to express their views effectively and mobilize support for them. The main advantage of the non-directive approach, then, is that it allows for greater participation in decisions relating to the definition of project goals, the organization of production, and the administration of social and economic affairs.

On the other hand, the directive approach is *also* frequently recommended for its organizational and economic advantages, although here one is thinking primarily of so-called 'national' or 'structural' benefits. It is said, for example, that centralized organization and administration of development projects enables Government to maximize output and efficiency and to programme future production in such a way as to accord with national priorities and needs. The maintenance of close administrative control is thus justified as being important for the achievement of overall economic objectives, and for contributing to political stability by preventing the emergence of so-called 'subversive' political groupings that might later pose a threat to the State and its policies.

An overview of Latin American situations suggests that the directive strategy is more likely to be favoured when the State is involved in large-scale commercially important investment projects associated with settlement and land development or land reform. A centralized mode of organization apparently allows for larger-scale extensive systems of cultivation and for the building of necessary infrastructure. It seems especially typical of two distinct types of development situation. The first involves the expropriation of private estates (*haciendas*) or company plantations producing important cash crops for national or international markets. Because these products are of vital importance to the national economy, either in terms of foreign exchange-earning capacity or the satisfaction of domestic needs, the maintenance of the pre-existing units of production is generally advocated in an attempt to preserve existing levels of production. Although new management structures, entailing increased participation by workers, and modifications in the patterns of work are often introduced, measures are taken to ensure that changes in the type of crops grown or in the pattern of capital investment do not jeopardize the existing system of production. In most cases, the State or agency retains the right to supervize certain administrative and financial matters, and to veto or intervene when its interests are threatened.

A striking example of this process is the Peruvian Agrarian Reform of 1969 carried out by President Velasco's military *junta*. As C. Collin Delavaud shows in chapter 3 of this volume, this reform included the expropriation of major sugar and cotton estates on the coast and the commercially important livestock *haciendas* of the central highlands. Although a cooperative mode of organization was introduced, existing units of production were kept intact, and technical and administrative personnel were given considerable control over the production process and over strategic socio-economic decisions. A similar pattern emerged during the post-Cárdenas period of the Mexican land reform when collective *ejidos*, producing commercial crops, lost much of their former autonomy and became tied to the State-organized Ejido Bank through its monopoly over credit and marketing: 'in a very real sense, the State became a new patron' (Hewitt de Alcántara, chapter 2 above). In Ecuador too, though on a lesser scale, we find evidence of land reform resulting in the establishment of State-controlled rice cooperatives in the Guayas basin of the coast (Redclift and Preston, chapter 4 above).

The second type of development situation is one in which the State is heavily involved in the development of basic infrastructure and natural resources for the settlement of virgin or underpopulated areas. Zones within these areas may be designated for cash-crop production. Hence, like the first situation, the State, wishing to ensure a high return on its investment, will again opt for what it evaluates as a low-risk strategy entailing minimum delegation of management decisions to the settlers and maximum control by State agencies of the means of production, technological and financial inputs, and marketing. Several examples of this are discussed in earlier chapters: the Mexican river basin development and tropical lowlands colonization schemes which, by and large, favoured collectivized forms of production and strong bureaucratic control (Revel-Mouroz, chapter 6 above); and the various ongoing irrigation, flood protection, and colonization schemes of coastal Ecuador made possible by revenue from oil and increased foreign investment (A. Collin Delavaud, chapter 5 above).

In both types of development situation, cooperative patterns of organization are commonly established. These, whilst appearing to offer a high degree of participation on the part of the beneficiaries or settlers, in fact frequently permit only limited autonomy of decision making, given the directive nature of the process and the major rôle played by Government agents in the design and operation of the production units. The cooperative idiom is used, it seems, to present the image of furthering participative democracy when Government actually intends to retain close control over the processes of economic and political change. In the Latin American context (with the exception of Cuba and Allende's Chile), the decision to opt for cooperative forms of organization and production has not, on the whole, been motivated by a desire to more towards the creation of a socialist society. On the contrary, the main motivating factors appear to have been the achievement of economic growth and the maintenance of political stability within the framework of a mixed, private and State-organized, capitalist economy (see Fals Borda, 1969, 1971; Long and Roberts, 1978). This is illustrated most clearly by the Peruvian Agrarian Reform. It occurred at a point when the national economy faced serious problems and when political movements among highland peasants and agricultural workers on the coast were at their height. Its main objectives were to achieve a more equitable distribution of productive resources, thereby stimulating commercial agriculture and a flow of capital into industry, and to establish new 'cooperative' and 'participative' forms of organization that would solve, or at any rate ameliorate, the tense political situation.

The few cases we have of Governments explicitly promoting cooperatives as part of a strategy of wealth redistribution, such as the Cárdenas land reform programme in Mexico during the 1930s or Allende's attempts in Chile in the 1960s and 1970s, suggest that this task is extremely difficult to achieve. According to Kay (1978, p. 128), the agrarian policy of Allende's Unidad Popular fundamentally failed because it focused too much attention on the middle and small-scale independent peasant sector and neglected to develop the revolutionary potential of the landless agricultural workers by strengthening collective forms of organization among them.

In most cases, then, a directive strategy has entailed the creation of State-supervised 'cooperative' institutions practising group farming, where water, credit, and sometimes land are allocated on a collective basis, but where the participation of members is severely limited. Although additional household plots may sometimes be provided, these are earmarked for subsistence cultivation and are segregated from the main areas of cash-crop production.

As I suggested, directive change is most clearly manifest in large-scale State-sponsored programmes of rural development. However, certain of its features are also present in situations where State agencies or private enterprise indirectly control the growing and processing of cash crops. For example, Luisa Paré (1977) shows how the *cañeros* (sugar-cane growers) of Atencingo, Mexico, use their own small *ejido* plots for sugar production, but are effectively organized and financed by nearby private or State-owned sugar mills. An important element in this system is the fact that Government legislation obliges all *ejidatarios* whose lands lie within a certain radius of the sugar mills to cultivate sugar and to rotate it with other crops when instructed to do so by the engineer in charge. Paré goes on to argue that this reproduction of the labour force for sugar production makes these peasant producers in many respects like an agricultural proletariat: in fact, at one point she describes them as 'proletarians disguised as peasants' (Paré, 1977, p. 51). Other cases of this type of close control of small-scale production are provided by Pebayle (chapter 7 above), who outlines the 'dictatorial' actions of the Souza Cruz (alias British–American Tobacco Company) vis-à-vis German colonists in southern Brazil, and by Feder (1978) in his analysis of the ways in which United States finance and technology determine the operations of the Mexican strawberry industry, as well as other types of export production (e.g. cacao, cotton, bananas, and tomatoes) in other parts of Latin America (chapter 13 above).

CONFLICTING INTERESTS AND ADAPTIVE STRATEGIES

A major obstacle, however, to the effective implementation of a policy of directive change, whether by the State, private enterprise or multinational company, is the existence of competing claims on resources by the different social classes and the conflicting interpretations of policy objectives emanating from the bureaucracy or agency itself. Any appraisal, therefore, of directive change (or, for that matter, any other rural development strategy) must start from the premise that it necessarily involves multiple actors and groups who differ in their interpretations, goals, and resources.

Given this, one can reformulate the argument somewhat by suggesting that under a directive strategy it is the State, by virtue of its direct control over major resources and its ability to impose sanctions or use coercive measures to achieve its goals, that constitutes the dominant actor, although this does not rule out the possibility of other actors influencing outcomes or changes of policy. Indeed, in most cases, the State avoids confrontation and will attempt to win the support of the members of the new settlements or cooperatives, if necessary by making certain concessions, providing these do not substantially reduce its own power base. This occurred, for example, with the collective *ejidos* of the Carrizo Valley of north-western Mexico studied by Winder (1979). The original plan had been to establish a fully collectivized system of agriculture. However, under pressure from the *ejidatarios*, the Government eventually gave way and allowed the organization of production to be subdivided into sectors and working groups; though, at the same time, it made provision for the protection of Government investment through retaining control over what crops were grown and how they were marketed. Government officials also turned a blind eye to internal arrangements for the distribution of profits on the basis of plot yields rather than labour input which had been the original principle adopted. They likewise agreed to abandon the idea of having professional managers heading each production cooperative when this was opposed by the members and permitted some of these functions to be carried out by locally-elected, peasant representatives.

It is important not to consider the State bureaucracy or sponsoring agency as undifferentiated in its interpretation of policy. Several studies (for example, Coombs and Ahmed, 1974, pp. 204–30; Grindle, 1979; Winder, 1979) bring out the complex ways in which different Government departments may adopt contrasting development priorities whilst justifying their actions and decisions by reference to the same policy statements. Thus, for example, banking and credit institutions often stress the need to maximize production levels to achieve optimum debt repayment, agencies responsible for water resources stress the rational utilization of irrigation supplies, and, in contrast to these aims which are essentially technocratic, ministries of land reform or their equivalent are likely to be more concerned with the achievement of social and organizational objectives, focusing on raising levels of participation and improving cooperative skills. In the final analysis, however, it is generally those agencies with greater budgetary and organizational resources and political support from the State executive that establish the predominant trends.

The issue of conflicting interests and goals is further complicated by the fact that, when making particular decisions, individual bureaucrats or international experts will be affected, either directly or indirectly, by their own personal career aspirations and existing networks of political alliance and obligation. Indeed, it has been suggested that in Latin America one of the most fruitful arenas for entrepreneurship, both political and economic, is the State bureaucracy itself (Glade, 1979).

When one shifts one's focus to consider the position of beneficiaries under a directive approach, one finds that, obviously, they are placed in a subordinate and dependent position *vis-à-vis* the State or sponsoring body. This is shown most clearly by the fact that title to, or full *de facto* control of, land and productive resources is very rarely vested in the peasant beneficiaries themselves, even under major land reform programmes. Furthermore, the latter are usually required to plant certain crops, use specific seeds, and adopt certain recommended methods of cultivation. Yet, despite these severe constraints on production behaviour, there is plenty of evidence that peasants do manage to evolve a series of tactics aimed at increasing their control over, for example, such aspects as the allocation of work tasks and the distribution of returns from production (Winder, 1979).

This adaptive process is not usually counter-productive for the State or agency, as it seldom threatens the smooth operation of production, nor does it affect the recovery of debts, and in fact it will probably tend to assist both. In addition, adopting a pragmatic approach to peasant and local-level responses to Government-designed systems of production has a useful payoff, in that political confrontation is thereby avoided. Also the furtherance of individual household-based strategies, which Webster (chapter 9 above) regards as reflecting 'shrewd initiatives based upon social, ecological, and technological realities', often leads to competition among the peasant beneficiaries themselves and hence reduces the possibility of the emergence of cohesive class-based political mobilization. This results in the reinforcement of social and economic differentiation amongst the peasantry, leading to small, but significant, differences in the rates of capital accumulation. Frequently the richer peasants diversify their economic interests into various non-agricultural activities, such as commerce, transport, or the processing of foodstuffs, thus reducing their dependence on the scheme itself. However the majority of peasants remain dependent clients of the State or agency and bear the major risks of production. Also, such organizations that exist, ostensibly to protect and further their interests, mostly function as systems of cooptation and control, since they are established by the sponsoring body and positions of responsibility within them are usually filled by the more influential and better-off peasants, whose interests often lie in perpetuating vertical patterns of control rather than in working to change them. A somewhat unusual example of the benefits to be gained by the monopolization of power positions within Government-sponsored production cooperatives is that described by Solano (1978) for a village in the central highlands of Peru. Solano documents how a group of founder members of the cooperative, of whom several were prominent in local politics, were eventually able to realize the capital invested in the original scheme to purchase their own *hacienda*, and how later a number of intragroup conflicts developed amongst them.

In addition to these major sets of actors, most cases of large-scale settlement or land reform schemes involve the indirect participation of other interest groups such as intermediaries, machinery contractors, transport entrepreneurs, political bosses, and agricultural union leaders, who use the presence of the scheme to advance their own particular political and/or economic objectives in local, regional, and sometimes national, arenas. Furthermore, projects which are located near major centres of population or which attract poor or landless immigrants generally face the problem of pressure from groups of non-beneficiaries who view the settlers as highly privileged and who struggle to obtain access to resources for themselves.

The changing patterns of competition and alliance between social classes and groups differently affected by programmes of rural development is a central issue for analysis in any study of directive change, since the presence of the State or sponsoring agency acts as a catalyst for wider regional-level changes. This process is for example, illustrated by C. Collin Delavaud's brief mention of the struggles that arose during the Peruvian Agrarian Reform between ex-*hacienda* workers, sharecroppers, and members of neighbouring peasant communities, and between Government agencies and certain agricultural unions largely controlled by left-wing groups (C. Collin Delavaud, chapter 4 above; see also Harding, 1974; Scott, 1972; for Mexican examples see Hewitt de Alcántara, 1976, chapter 2 above; Winder, 1979).

Feder (chapter 13) extends the analysis to consider the cumulative effects of massive transfers of United States capital and technology organized by United States agri-businesses and investment firms. This process, he suggests, is having major consequences for the rural inhabitants of Latin America, resulting in a rapid concentration of control over land, productive resources, and agriculturally-related industries and services, a widening of income differentials, and increased proletarianization and rural depopulation. Not surprisingly this pattern has coincided with increased emphasis on large-scale directive strategies involving the use of sophisticated labour-saving machinery and equipment and modern farm management methods.

This mosaic of competing and, at times, overlapping interests presents the State and its officials with the major problem of reconciling these differing needs, whilst at the same time not wishing to deviate substantially from accepted policy. Systematic evaluation of the directive strategy therefore requires that one identify the particular groups of actors affected and their respective interests, document how they interact, and attempt some assessment of the benefits or gains they achieve. It is not possible in a strict sense to arrive at an aggregated view of the costs and benefits of a particular development programme since the existence of divergent interests entails that benefits to one set of participants may well constitute important costs for others. In addition, it is important to place this interplay of interest groups against broader economic and political parameters. Hence, project evaluation necessitates a careful analysis of the distribution of power between the various parties and an understanding of the politics of resource allocation within a regional and national context. This viewpoint concurs with Hewitt de Alcántara's (chapter 2 above) insistence that agrarian reform implies 'not only a redistribution of land, but also a redistribution of power'.

My earlier consideration of adaptive strategies which modify and introduce new elements, raises the question of the likely reactions of beneficiaries or settlers to circumstances of economic reorganization such as might result from the collapse and later revival of a particular export crop, or from the withdrawal of State assistance during times of economic recession. Several chapters in this volume provide insights into this process, showing how peasant farmers and agricultural labourers build upon existing patterns of social organization, often based on household enterprise, to devise new livelihood strategies. We find, for example, in Ecuador that, once cacao ceased to be a profitable export for coastal plantations, the former tenants (*finqueros*) were able to take over the estates for the cultivation of food crops, especially rice. Then later, when demand for cacao returned and they had consolidated their position, the same *finqueros* assumed a major rôle in the organization of cacao production and, through the recruitment of sharecroppers (*desmonteros*), in the cultivation of rice (A Collin Delavaud, chapter 5 above). A similar example is given by Pebayle (chapter 7 above) in his discussion of post-pioneer colonization in southern Brazil, although he argues that it is the big *fazeindeiro* planter rather than the smallholder who is more easily able to switch crop or type of agriculture. His discussion also draws attention to the ways in which different ethnic groups come to specialize in the production of different products; the Mennonites concentrating on dairying, the Italians on wine, and the Japanese becoming major producers of vegetables and poultry and forming their own highly efficient cooperatives.

The significance of the ethnic factor for analysing differential responses to development is further emphasized by Arizpe (chapter 8 above) and Webster (chapter 9 above). Arizpe provides detailed evidence for the evolution of different socio-economic strategies and levels of economic performance among indians and mestizos in the central part of Mexico, despite the fact that each group benefited equally under the land reform programme. Her argument includes not only an account of rural social organization but also describes the different migration experiences of the two groups. Throughout, she stresses how cultural orientations and attitudes, for example concerning education and urban life, interact with differential access to employment and economic opportunities to reinforce differences of ethnic and socio-economic status. She also discusses the process of cultural centralization which has accompanied political centralization in Mexico and suggests that this is leading to certain forms of cultural alienation amongst peasant and poor urban classes.

Webster's analysis likewise focuses on ethnicity, arguing against simplistic class interpretations of rural social structure in highland Peru. One implication of his argument is that major programmes of State-organized development, such as land reform, are seriously handicapped if they fail to recognize the integrity and development potential of ethnically organized peasant groupings. According to Webster, the bureaucratic and paternalistic nature of rural development policy in Peru works against the achievement of its essential goals. 'Policy', he claims, 'must humbly adapt to the plural reality of Peruvian society' (Webster, chapter 9 above).

Not all these examples refer to development situations where a directive strategy is clearly in force, but they do underline the importance of exploring the consequences of particular policies through an understanding of the differentiated and heterogeneous nature of Latin American rural society. Most regions of Latin America have been profoundly affected by colonial and capitalist development. In most rural areas, we are already dealing with a 'transformed' peasantry—one which has experienced major economic change and planned intervention by the State. Central to one's analysis, then, is the isolation of factors responsible for variations in the economic and political strategies pursued by different types of household and social group. Such an approach requires, as Arizpe and Webster show, the inclusion of cultural factors, such as ethnicity and kinship ideology and sentiments, as well as economic incentives and constraints.

A further related point is that the study of development should basically be concerned with social *actors*, not just *structures*. It is individuals and networks of interacting individuals—i.e. status groups and social classes—who attempt to give meaning to the world around them and who organize ways of dealing with it in accordance with their own felt needs and interests. Hence, although the context of action is shaped by structures external to the individual (e.g. the state of the market, the distribution of wealth in the society, and the type of political régime), it is important to explore how particular households and social groups (what Adams (1975) calls the 'operating units' of society) respond to certain types of State intervention and change.

Such a theoretical perspective on development does not, I believe, neglect the structural determinants of action or the possible unintended consequences of policy or behaviour. Indeed, even in the most closely controlled development programmes, one must allow for unanticipated and unintended outcomes that result from the interplay of various environmental factors and from the struggles that occur between competing social classes. Rural development programmes oriented, for example, towards the intensification of agricultural production so as to improve the subsistence base and increase the marketable surplus of peasant farmers have often had the opposite effect of displacing large numbers of the peasant population. In fact, as several earlier chapters indicate, improved forms of agriculture frequently lead to land concentration in the hands of an emergent entrepreneurial class, the creation of a landless or poor peasant labouring class, and increased out-migration to urban centres. These processes must be analysed not only in terms of the ways in which Government policy or the interests of dominant classes structure the development pattern but also through a detailed examination of the characteristics and resources of all relevant actors and social groups. Hence one needs to look closely at the client population in order to determine which groups benefit most from the specific policy implemented and how this happens. And it is equally essential to enquire into the operations of the bureaucracy and to examine the interface between the local agencies responsible for interpreting and implementing policy and the peasant population who reinterprets it in accordance with its own interests.

A FINAL COMMENT ON 'PARTICIPATION'

The use of an actor-oriented approach in the study of directive change in Latin America has implications for how one interprets and studies the question of 'participation'. A major difficulty when discussing participation is the tendency to assume that it is something that can be measured in some objective or absolute sense. Political scientists, such as Verba et al. (1973), attempt to do this cross-nationally by measuring such activities as voting, campaigning, contacting officials, and being involved in community organizations. However, as Bayliss (1976) has argued, 'participation' has many diverse meanings in as many different settings. Studies that aim to measure the quantity, distribution and correlates of 'participation' among individuals within and between societies overlook the fact that it is the context, functions, and subjective meanings of the concept, as they relate to various social actors under particular political frameworks, that is the critical issue. Thus, according to Bayliss, one should differentiate between an élite's use of 'participation' aimed at establishing and maintaining the legitimacy and stability of the political unit which they control, and the ordinary citizen's or peasant's interpretation in terms of influencing Government (although the expectation of success may be low) or as a means of acquiring certain benefits for the individual or group. Also, whether or not a particular power structure encourages definitions in terms of influence and how much 'ceremonial participation' there is, must be determined by analysis, not used as a yardstick of an evaluative kind.

Perhaps, as Adams (1976) has forcefully argued, 'participation' is best seen as a two-way or, as I would prefer, multichannelled competition between people and Government (i.e. between different social classes) that is primarily composed of behaviours designed to realize the best interests of the participants through immobilizing the efforts of those whose interests are seen to be putatively contrary. Hence, 'participation' is not simply a positive effort by citizens or a ruling class to further its interests: it is equally oriented towards obstructing or inhibiting the realization of the contrary interests of others. Participation, therefore, is inevitably linked to the operation of power structures and to the question of how the State attempts to create constraints on the participation of specific social sectors, particularly those that are seen to pose a threat to its policies or political legitimacy.

These dimensions, I believe, are central to an understanding of the increased emphasis on directive change in Latin America, since such a strategy reduces local-level participation to a minimum. But, as I have also stressed, constraints on action can never be totally effective—contradictions do arise, adaptive strategies are evolved, concessions are made, and outcomes are frequently unintended.

BIBLIOGRAPHY

Adams, R. N., 1975, *Energy and Structure, A Theory of Social Power* (Austin and London: University of Texas Press).
—— 1976, 'Power and Participation in Latin America'; Paper presented to *Seminar on Faces of Participation in Latin America*, University of Texas at San Antonio, November.
Batten, T. R., 1974, 'The major issues and future direction of community development', *Community Development Journal*, 9, No. 2.
Bayliss, T., 1976, 'The meaning of participation: the ideological and instrumental aspects'; Paper presented to *Seminar on Faces of Participation in Latin America*, University of Texas at San Antonio, November.
Chambers, R., 1969, *Settlement Schemes in Tropical Africa* (London: Routledge and Kegan Paul).
Coombs, P. H., and Ahmed, M., 1974, *Attacking Rural Poverty: How Non-Formal Education Can Help* (Baltimore and London: The Johns Hopkins University Press, for International Bank for Reconstruction and Development).
Fals Borda, O., (Ed.), 1969 *Estudios de la Realidad Campesina: Cooperación y cambio*, Vol. 2 (Geneva: United Nations Reaserch Institute for Social Development).
—— 1971, *Cooperatives and Rural Development in Latin America: An Analytic Report*, Vol. 3 (Geneva: United Nations Research Institute for Social Development).
Feder, E., 1978, *Strawberry Imperialism* (Mexico City: Editorial Campesina).
Glade, W. P., 1979, 'Decision-making in the entrepreneurial State'; in Strickon, A., and Greenfield, S. (Eds), *Entrepreneurs in Cultural Context* (Albuquerque: University of New Mexico Press).
Grindle, M. S., 1979, *Bureaucrats, Politicians and Peasants in Mexico* (Los Angeles: University of California Press).
Harding, C., (1974), *Agrarian Reform and Agrarian Struggles in Peru*, Working Paper, No. 15, Latin American Centre, Cambridge University.
Hewitt de Alcántara, C., 1974, *La Modernización de la Agricultura Mexicana, 1940–1970* (Mexico City: Siglo XXl).
Kay, C., 1978, 'Agrarian reform and the class struggle in Chile', *Latin American Perspectives*, V, summer.
Lehmann, D., 1974 *Agrarian Reform and Agrarian Reformism* (London: Faber and Faber).

Long, N., 1977, *An Introduction to the Sociology of Rural Development* (London: Tavistock Publications Ltd).

Long, N., and Roberts, B. R. (Eds), 1978, *Peasant Cooperation and Capitalist Expansion in Central Peru* (Austin: Institute of Latin American Studies, University of Texas Press).

Long, N., and Winder, D., 1979 'The limitations of "directive change" in rural development: examples from Latin America'; Paper presented at *Xth Congress of the European Society for Rural Sociology*, Córdoba, Spain.

Paré, L., 1977, *El Proletariado Agrícola en Mexico, Campesinos Sin Tierra o Proletarios Agricolos?* (Mexico, Madrid, and Bogota: Siglo Veintiuno).

Scott, C., 1972, 'Agrarian reform, accumulation and the role of the State: The case of Peru'; in *Dépendence et Structure de Classes en Amérique Latine*, **IV**, Seminaire Latino Americain, CETIM (AFJK), October.

Solano, S. J., 1978, 'From cooperative to hacienda: The case of the agrarian society of Pucará'; in Long, N., and Roberts, B. R. (Eds), *Peasant Cooperation and Capitalist Expansion in Central Peru* (Austin: University of Texas Press).

Verba, S., Nie, N. H., et al., 1973, 'The modes of participation: Continuities in research', *Comparative Political Studies*, **6**, July.

Winder, D., 1979, 'An analysis of the consequences of Government attempts to promote community development through the creation of cooperative institutions, with special reference to rural Mexico', *PhD Thesis*, Manchester University.

Author Index

Adams, 246
Arizpe, 189, 203
Azuaras Salas, 99

Bataillon, 189, 191
Batten, 237
Bayliss, 246
Beltran, 98
Bencomo, 84
Benitez, 97
Bourricaud, 136
Bonfil, 124
Bourque, 152
Brown, 218, 225
Browning, 12

Camacho, 83, 84
Celestino, 178
CENETAL, 98
Chambers, 236
Chenery, 227 *et seq.*
Colby, 150
Conklin, 152
Coombs, 236, 241
Crane, 203

Della Paolera, 182
Dozier, 6

Enjalbert, 83

Fals Borda, 239
Favre, 8
Flores, 141, 148, 152
Fort, 99
Foster, 142
Frank, 146
Freire, 16

Gambaccini, 182

Gaude, 203
Germani, 161
Glade, 242
Goldberg, 214
Gow, 147, 152
Greaves, 146
Griffin, 201
Grindle, 241

Handlemen, 152
Haney, 198
Harding, 152

Kay, 240
Kemper, 240
Kleinpenning, 7

Lehmann, 236
Lerner, 124
Long, 239

Margulis, 112
Martinez Alier, 63, 140, 148, 152
Monbeig, 112
Murra, 8

Nagata, 150

Paré, 240
Philpott, 203
Pitt-Rivers, 12
Port-Levet, 97
Primov, 135, 148, 150

de Queiroz, 118

Randle, 182
Redfield, 124
Ribeiro, 124
Rivière d'Arc, 7
Roberts, 7, 124, 239

Rofman, 182
Rogers, 198, 201

Salomon, 8
Singer, 124
Solano, 242
Sollis, 203
Stavenhagen, 124, 146
Stewart, 8

Tabah, 191
Taveras, 61, 62

Unikel, 189

Van de Berghe, 124, 126, 135, 147, 148, 159
Vasconcelos, 127
Velasco Suárez, 97
Verba, 246

Wilkie, 182
Winder, 241, 242
Wolf, 124, 226

Zuidema, 153
Zuvekas, 57

Subject Index

Abancay, 44, 45, 49
Acapulco, 101
Advance to the Sea, 83, 85, 88, 100
African oil palm, 78
Agrarian bureaucracy, 31
Agrarian change, 199
Agrarian Codes, 2, 6, 84
Agrarian reform, 2, 21, 34, 44, 45, 58, 62, 135, 186, 213
  Laws, 54, 74, 79
Agricultural enterprise, 35
Agriculture, perceptions of, 206
Agriculture-related industries, 223
Aid, 16, 17, 57
Alausi, 196
Aleman, M., 84
Aleman Dam, 94
Alcavitoria, 148
Allende, S., 240
Alliance for Progress, 54
Altos de Chiapas, 97
Amazonia, 7, 44, 103, 104, 118
Ancash, 49, 136
Andahuaylas, 49
Angostura Dam, 94, 100
Animation rurale, 236
APRA, 37, 38, 46
Apurimac, 49, 147
Arabs, 56
Aracatuba, 114
Araraquara, 113
Arequipa, 47, 51
Asians, 56
Atahualpa, 60, 200, 207
Atencingo, 240
Ayacon (Hda), 62
Ayacucho, 47, 49, 148
Azuay, 61

Babahoyo, 68

Babahoyo Pilot Project, 76
Baja California, 91
Balancán–Tenosique Plan, 99
Balsas Commission, 88
Balsa wood, 68
Bananas, 68
Banco Comercial de Guayaquil, 54
Banco de Crédito Peninsular, 99
Banco de Crédito Rural, 96
Bandeirantes, 103
Banking institutions, 241
Bank of Brazil, 113, 116, 117, 118
Bauru, 113
Belaúnde, F., 44
Belize, 99
Blumenau, 110, 111
Bolivia, 2, 9
Bonampak–Palenque road, 98
Brazil, 8, 220
British–American Tobacco Company, 115
British foreign policy, 17
Buenos Aires, 13, 157, 158, 159 *et seq.*, 170, 179

Cacao, 77
Cacapava, 117
Cactus, 100
Caiua, 113
Cajamarca, 41, 47
Campeche, 99, 100
Campo Grande, 117
Cancun, 100
Capital, foreign, 219
Cárdenas, L., 25, 26, 27, 28, 33, 35, 84, 239, 240
Cargo system, 26
Caribbean, 195, 203, 207
Carriso valley, 241
Carrizal–Chone canal, 76

251

Canchagua, 200, 202
Cangrejera, 100
Cantinflas, 129
Catamarca, 163
Cavendish bananas, 69, 77
Caxias do Sul, 110, 111
CCI, 93
Ceja de la Montana, 42, 136
Cerro de Oro dam, 96
CGIAR, 218
Chaco, 163, 170
Chancay, 41
Chapecó, 110
Cheqec, 137, 141
Chiapas, 83, 84, 85, 140, 148, 185
Chile, 157, 239, 240
Chimborazo, 196
Chinese, 56
Choapas, Las, 100
Chols, 97, 98
Chone, 68
Chontalpa Plan, 92, 93
Ch'ilka, 137
Christianat, 147
Chubut, 172
Ch'uchu, 137
Church, Catholic, 17, 59, 145, 200
Chumbivilcas, 148
CNC, 91
Coahuila, 91
Coast, 39
Coatzacoalcos, 100
Cobaay, 91
Cocoa estates, 56, 68, 79
Coffee growing, 111
  programme, 78
Colima, 88
Colombia, 7
Colonization, 6, 204, 239
  Collective *ejido*, 91
  Laws, 84
  market-oriented, 8
  National Plan of Ejidal, 89
  planned, 69, 80, 100
  spontaneous, 6, 67, 69, 76, 80, 97, 98, 100
Commodity systems, 211, 225
Communal land, 23
Communications, 31
Community development, 236
Communities
  indian, 43, 47, 54, 136

integration of, 49
peasant, 243
production on, 243
CONASUPO, 187
Concheros, 129
Cooperatives, 49–50, 58, 239
Co-parenthood, 141
Cordoba, 158, 163, 170
CORFONOR, 77
Corozal, 98
Corrientes, 163
Conservation, soil, 206
Constitution of 1917, 84
Convención, La, 42
Cotton programme, 74
Counsellor, financial and technical, 107
Craftsmen, 188
Crops
  commercial, 32
  export, 73, 74
Cruz Alta, 117
Cuba, 35, 193, 213, 239
Culpulli, 25
Cultural change, 123
Curitiba, 110
Cuzco, 42, 49, 135, 136, 138, 148, 149, 151
Cycles, economic, 54

Dairying, 114
Daule, 68
Daule-Peripa dam, 75
Debt, 32
Decreto 1001, 55, 66
Del Río, Dolores, 132
Depopulation, 222, 226
Development assistance, 217
  programmes, 74
Diamante, 182
Díaz Ordaz, G., 89
Díaz, Porfirio, 84
Directed change, 236, 237
Dotejiare, 130
Dourados, 117

Echeverría, L., 34, 35, 89, 91, 94, 99
Ecological zones, major, 39
Economy, dual, 53
Ecuador, 12, 14, 57, 98
Education, non-formal, 16
Edzna Project, 91
Ejido, 4, 25, 185, 291

SUBJECT INDEX

Ejido Credit Bank, 25, 28, 29
El Oro, 68, 71, 72
El Salvador, 12
El Triunfo, 72
Emigrant labour, 172
Emigration
 causes of, 14
 characteristics, 11
 rural, 10
Entrepreneurship, 242
Entre Ríos, 13, 158, 172, 182
Esmeraldas, 68, 71, 72, 75, 80
Estates, 53, 238
 expropriation of, 59
 large, 47, 54
 sale of, 61
 stock-raising, 43
Ethnicity, 9, 123, 135

Farmers
 middle-sized, 39
 tenant, 43
Felix, María, 132
Fomeque, 198
Foreign investment, 27
Forest, clearing of, 97–8
Formosa, 163, 170
Fotonovelas, 132

German dialect, 172
Gomez Pompa, 96
Granada Television, 9
Gros Michel bananas, 69, 78
Guachapala, 200
Guamote, 60, 61
Guanajuato, 100, 188
Guaranda, 196
Guatemala, 9, 126, 148
Guatemalan frontier, 99
Guayaquil, 54, 69, 72, 75, 80
Guayas, 68, 71, 72, 76, 79, 80
Guayas Basin, 5, 55, 75, 78, 239
Guyana, 99

Haciendas, 41, 47, 54
Handicrafts, 126
Hemp, 78
Highways, 75
Holdings, large private, 27
Huancavelica, 148
Huanto, 47
Huaquillas, 75

huasipungo, 53, 55, 59, 61, 62
Hydroelectricity, 100

IBGE, 110
Ica, 42, 51
IDS, 227
IERAC, 59, 74
Ijuí, 110
Indian areas, 199
Indians, forest, 9
Industries
 agriculture-related, 223
 cottage, 13, 16
 dairy, 14
 small rural, 32
Infrastructure, 75
INIAP, 74
Innovation, agricultural, 109, 201
Instituto Nacional Indigenista, 127
International lending agencies, 221, 236
Ireland, 195
Italian areas, 115

Jalisco, 188
Japanese immigrants, 115
Jequetepeque, 41
Joinville, 110
Jubones valley, 68
Jujuy, 163, 170
Jundiaí, 115
Junín, 47, 49, 51
Junta de Asistencia Social, 59

Kaykay, 148, 152

Lacandon forest, 97
 indians, 97, 98
Laguna, La, 25, 99
Lambayeque, 46, 47, 51
Landless labourers, 25
Landowners, large, 24, 79
Land reform, *see* Agrarian reform
Land Sale Guaranty Plan, 51
Latifundio, 59, 215, 223,
 reconstructed, 35
Leche valley, 41
Lerma Plan, 187
Libertad, 40
Limones, 75
Loja, 61
Lopez Mateos, 88
Lopez Portillo, 91, 92

Los Rios, 72, 79, 80, 99

Machala, 72
Majes canyon, 41
Malpaso dam, 82
Manabí, 68, 71, 78, 80, 196
Mariategui, C., 37
Marília, 113
Maringá, 113
Market economy, 141
Market gardening, 114, 115
Market penetration, 53
Market pressures, 54
Market relations, 54
Marqués de Gomilla National Park, 49
Mass media, 135
Matlatzinca, 126
Mato Grosso, 106
Mazahua, 10, 11, 12, 123, 125, 203
Mechanization, 222
Mendoza, 170, 173
Mennonites, 115, 244
Merced, La, 128, 189
Mestizos, 72, 125, 127, 128
Mexican Irrigation Districts, 29
Mexican River Basin development, 239
Mexico, 2, 3, 7, 8, 11, 13, 159, 222, 224, 240
Mexico, State of, 188, 190
Mexico City, 13, 85, 128, 132, 189
Mexico—US border, 125, 128, 192
Michoacán, 100
Migrants, 198
Migration, 128
  attitudes to, 190
  reasons for, 175
  return, 12, 175, 204
  seasonal, 56
Milagro, 68
Milagro Project, 76
Military government, 56
Minatitlán, 100, 101
Minifundista, 30, 42
Misiones, 163
Misiones culturales, 132
Misti, 137
Mobility, inter-ethnic, 15
Modernization, 1, 74, 123, 226
Monopolization of land, 223
Multinational companies, 236

Naranjal, 72

National Peasant Confederation, 25
National Plan of Ejidal Colonization, 89
Neuquén, 170, 172
New villages, 93
Non-directive approaches, 237
Nono, 60

Oaxaca, 83, 85, 94
Obregon, Alvaro, 99
Oca, 199
Ocongate, 151
Ocosingo, 97, 98
Overseas Development Ministry, 17
Odría, Manuel, 38
OIDE, 77
Oligarchy, 37
Ollachea, 147
Olmos, 41
Ona, 10
Oriente, 81
Otavalo, 149, 152
Otomi, 126, 129
Ourinhos, 113
Overseas agricultural ventures, 214

Palenque, 97
Pampa, La, 111
Panama disease, 69
Papaloapan Commission, 94, 95
Paraguay, 110
Paraná, 106, 108, 112, 117, 182
Paratia, 148
Participation, 246
Passo Fundo, 117
Pastoralism, 138
Pato Branco, 110
Pátzcuaro, 190
Paz, La, 151
Peasants
  landless, 46
  militia, 24
  organization, 21
Peladito, subculture of, 129
Pelotas, 115
Pena Colorada, 100
Penetration, external, 15–18
Perception of agriculture, 206
Permon indians, 99
Peru, 3, 4, 9, 10, 198, 207, 238
Petroleum resources, 73
Pichincha, 60, 61

SUBJECT INDEX

Pioneer mentality, 106
Piura, 40, 42, 46, 47, 51
Plantation economy, 53
Plantations, 39, 40, 43, 44, 45, 50, 68, 79, 236
Plata, La, 158
Poechas dam, 41
Polish migrants, 108
Polyculture, commercial, 93
Ponta Grossa, 117
Population growth, 32
Population pressure, 180
Porto Alegre, 120
Poza Honda dam, 76
Poza Rica, 101
Presidente Prudente, 114
Production co-operatives, 242
Programa Nacional de Desmonte, 89
Providencia, La, 125
Puebla, 188
Puna, 41, 136
Puno, 49, 148
Puerto Bolivar, 68
Puerto Maldonado, 138

Q'ero, 137, 149
Qeshwa, 137
Qolla, 137
Quechua, 149
Quetzaltenango, 148, 152
Quevedo, 68, 69, 72, 75, 78
Quilanga, 206
Quince Mil, 138
Quinindé, 72
Quinoa, 199
Quispicanchis, 147
Quito, 75, 80, 202, 205, 206,

Radio, 143, 204
Redistribution of income, 226
Remittances, 203
Revolution of 1910, 127
Ribiero Preto, 113
Rice, 56, 57, 68, 115
  cooperatives, 239
Riobamba, 59
Rio Grande do Sul, 108
Rio Grijalva Commission, 92
Rio Grijalva Project, 94, 99
Rio Hondo, 99
Rioja, La, 161, 163
Rio Negro, 172

Rio Obispo, 94
River Basin Commissions, 89
Rosario, 158
Runa, 137

Sale of land, 60
Salina Cruz, 100
Salta, 163, 170
San Francisco, 130
San Lorenzo, 40, 41
San Luis, 163
San Martín, 42
Santa Catarina, 106, 107, 108, 110
Santa Cruz, 172
Santa Elena peninsula, 68
Santa Fé, 41, 163, 182
Santa María, 115
Santa Rosa, 110
Santa Teresita, 60
Santiago de Chile, 163
Santiago Toxi, 130
Santo Angelo, 117
Santo Domingo de los Colorados, 55, 69, 72, 78
Sao Borja, 117
Sao Jose do Rio Prêto, 113
Sao Paulo, 7, 111, 157
Sao Roque, 115
Schools, 130, 143
Scotland, 195
Self-sufficiency, 31
Selva, 42
Sesmarias, 104
Shanty towns, 129
Sharecropping, 43, 79, 202, 243, 244
Sicuani, 138
Sierra, 41, 47, 58
Sinaloa–Sonora, 88
SINAMOS, 46, 49
Smallholders, 39, 53, 114, 223
Soccer, 143
Social change, 9
Social institutions, 141
Socialist empire, 135
Social organization, 138
Souza Cruz Company, 115, 240
Soviet Union, 17
Soya, 113
State authority, 21
State farms, 5
Strawberry industry, 240
Subsistence, 32

Superprofits, 224

Tabasco, 99, 100
Tarascan indians, 188, 191
Teachers, 204
Technology
 agrarian, 188
 foreign, 220
 imported, 17
 transfers, 217
Tehuantepec peninsula, 96, 185
Telenovelas, 6, 132
Television, 132
Tenancies, decline of, 60
Tierra del Fuego, 172
Tinajones, 4
Titicaca, Lake, 136
Tocra, 149
Tourist complexes, 100
Trans-Amazonian Highway, 109
Tupac Amaru, 37, 51
Tumbes, 41
Tuxtla–Pichucalco road, 97
Tzeltal, 97, 98
Tzotzil, 8

Ucum, 91
United Fruit Company, 220
United Kingdom, 15
United States, 11, 13, 190, 211, 220, 221, 224
Urban growth, 72, 107

Urban hierarchy, 159
Uru, 10
Uruguay, 157
Usumacinta, 97, 98, 99
Uxpanapa Plan, 94, 95, 96

Varayoq, 144–145
Velasco, A., J., 238
Velasco, I., J.M., 54, 57
Venezuela, 2, 99, 158
Villahermosa, 100
Vinces, 68
Volga Deutsch, 172 *et seq.*

West Germany, 223
Wheat, 113
Whites, 125
Wholesalers, 107
Witwarsum, 115
World Bank, 221, 227

Xcan, 91
Xochimilco, 128

Yaqui valley, 25
Yucatan, 29, 83, 84

Zacatón, 126
Zapata, 3, 23
Zapotlango, 188
Zarate, 182